FLUID POWER MAINTENANCE BASICS AND TROUBLESHOOTING

FLUID POWER AND CONTROL

A Series of Textbooks and Reference Books

CONSULTING EDITOR

Frank Yeaple

President
TEF Engineering
Allendale, New Jersey

ADDITIONAL VOLUMES IN PREPARATION

FLUID POWER MAINTENANCE BASICS AND TROUBLESHOOTING

RICHARD J. MITCHELL
Vickers, Inc.
Detroit, Michigan

JOHN J. PIPPENGER
Parker Hannifin
Cleveland, Ohio

CRC Press
Taylor & Francis Group
Boca Raton London New York

CRC Press is an imprint of the
Taylor & Francis Group, an **informa** business

CRC Press
Taylor & Francis Group
6000 Broken Sound Parkway NW, Suite 300
Boca Raton, FL 33487-2742

First issued in paperback 2019

© 1997 by Taylor & Francis Group, LLC
CRC Press is an imprint of Taylor & Francis Group, an Informa business

No claim to original U.S. Government works

ISBN-13: 978-0-8247-9833-8 (hbk)
ISBN-13: 978-0-367-40107-8 (pbk)

Library of Congress Cataloging-in-Publication Data

Mitchell, Richard J.
    Fluid power maintenance basics and troubleshooting / Richard J.
Mitchell, John J. Pippenger.
        p.   cm. — (Fluid power and control ; 14)
    Includes index.
    ISBN 0-8247-9833-3 (hc : alk. paper)
    1. Fluid power technology. 2. Oil hydraulic machinery—
-Maintenance and repair. I. Pippenger, John J. II. Title.
III. Series.
TJ843.M58 1997
620.1'06—dc21   97-56
                                                                  CIP

**Visit the Taylor & Francis Web site at
http://www.taylorandfrancis.com**

**and the CRC Press Web site at
http://www.crcpress.com**

Preface

The design engineer, machine manufacturer, technician, mechanic, and machine operator all recognize that there is no magic associated with troubleshooting fluid power systems. Fluid power transmission systems deal with predictable force and motion. Trouble may be broadly categorized as a *malfunction*: the machine does not perform in the usual programmed mode. To identify the change from normal to abnormal performance, those responsible for continued trouble-free operation must understand the normal functional operation so that they can identify those processes associated with force and/or motion that are functioning abnormally or have ceased to function.

All the basic troubleshooting essentials are addressed in this volume:

Communications. This includes circuit drawings, the bill of material, and the operational manual. Availability of these items is usually the responsibility of management. Explanations of how to use these essential materials are addressed.

Comprehension. The ability to read and understand symbols and the associated circuitry is the responsibility of the technician. This volume teaches fluid power symbology and the associated circuits.

Component knowledge. This, too, is the responsibility of the technician. Operation and care of major components is explained in this volume. Potential problems with components are addressed in each section.

Mechanical motions. An understanding of mechanical motions can be acquired by watching normal machine operation, questioning the operator about noted operational changes, and studying descriptive information in the operator's manual. Basics of many machine operations are covered to provide a logical starting point to assess service and/or troubleshooting requirements.

Troubleshooting. This is the summation of all of these factors: to find the cause of abnormal performance and apply the remedy. The availability of spare parts, maintenance and repair materials, appropriate tools, and fluid conditioning equipment needed to minimize the associated downtime is discussed.

The primary audience of this book is the engineering community, which must be aware of potential problems, as well as the mechanics and machine operators who are charged with operating and maintaining high performance machinery in critical and time-sensitive situations.

The first chapter discusses the symbology needed to read and comprehend the hydraulic circuit drawings and the flow paths involved. Symbolic materials are also available from the International Standards Organization (ISO), but mechanics and machine operators generally do not have ready access to this type of material. Chapters 2 and 3 are critical to basic troubleshooting functions. Illustrations were chosen to show function and construction in cut sections and expanded views to illustrate the basic operation of each component.

Once users learn to communicate and read circuit drawings, they can be introduced to the components involved in the power transmission system. After readers become acquainted with the components, they can move on to the primary basic circuits used in typical fluid power systems, such as material handling, prefill-type press circuits, typical injection molding machine circuits, construction machinery, utility maintenance machinery, fire-fighting equipment, and preventive maintenance information, which are important for any piece of hydraulically actuated equipment.

The typical heavy equipment manual initially provided with machinery is rarely available to those who need it most. This book provides every mechanic and machine operator with the troubleshooting essentials they will use in their daily work with fluid power equipment.

Richard J. Mitchell
John J. Pippenger

Contents

v

1

Basics

Technical knowledge with mental discipline is the key to logical, efficient trouble-shooting.

Improper troubleshooting procedures often result in unnecessary and costly hydraulic repairs—wasting time, money, and productivity. All too frequently the individual responsible for troubleshooting a machine hastily and arbitrarily decides that the problem is a faulty pump. Not unexpectedly, after the pump is replaced, the system still will not work.

An organized troubleshooting approach can eliminate such wheel-spinning, but for such a program to be effective, it is essential that personnel be fully aware that:

- leadership and direction originate with top management, and
- the objectives are lower maintenance costs and improved productivity through reduced downtime.

MENTAL DISCIPLINE

Effective and productive troubleshooting is not a complicated process. Clearly, the first prerequisite is technical knowledge, but equally important is mental attitude. In addition to understanding machine operation as well as hydraulic components and circuitry, the individual must also have the training and discipline necessary to troubleshoot a hydraulic system in a *logical and orderly* manner.

Quick and efficient hydraulic troubleshooting requires that workers follow logical steps to locate malfunctions in the shortest possible time. The challenge in resolving this problem consists of two steps:

1. Think through and identify the elements that might contribute to a system malfunction.
2. Avoid jumping to incorrect and illogical conclusions which may result in improper and inappropriate actions.

1

BACKGROUND

Within any company, troubleshooting personnel have different levels of competence, manual skills, and experience. For this reason, it is absolutely necessary that before individuals are assigned the task of troubleshooting hydraulic systems, they become *intimately* familiar with:

- hydraulic terminology
- standard graphic symbols
- component function and operation
- circuit diagram interpretation
- machine and system operation, and
- malfunction troubleshooting procedures.

If troubleshooters are not fully conversant with these elements, any program is likely to fail.

TERMINOLOGY AND SYMBOLS

A first essential is to be completely comfortable with fluid power symbology. The remainder of this chapter provides an introduction to symbology for the entry level mechanic and a quick reference for the designated troubleshooter.

The fluid power industry has a marginal, somewhat cloudy history of attempting to minimize piping and to reduce the size of the control components.

While it is true that early machine tool feed and traverse systems were compact and much piping was eliminated by designing many functional valves into a single casting, there was not one but two Achilles' heel problems. The first problem was the lack of standardized symbols for circuit analysis and a general lack of communication between designer, builder, user, and maintenance personnel. The second problem was the curse of all valves sharing a common housing: if one valve or a section of a valve within the casting assembly was damaged by poor quality machine maintenance or some accidental cause the entire valve structure was lost. Some of the units were quite costly and the attempts to keep them operational was often at the expense of unfixed leaks, marginal operation, and/or further contamination of the system. Some of these systems in well-maintained shops may still be in service.

International Standards Organization (ISO) symbols (*ISO 1219*) and the vastly expanded network of technical education encompassing fluid power is the approach to an answer for the first problem.

An approach to the problem of communications is suggested because we now have good, practical symbology which has worldwide acceptance. We also have many useful standards promoted by trade associations like the National Fluid Power Association and standards bodies like The American National Standards Institute (ANSI) and the International Standards Organization. Many international engineering societies and trade associations join to maintain and expand these fluid power communication channels.

A major problem which is being addressed in several areas is that of predicting the performance of fluid power systems and components. This is an ongoing activity that can respond to the communication systems now available to the designer, such as the computer information banks and new software recently entering the marketplace, which is capable of defining functional operation of components and ultimately the integration of the components into dynamic systems. Most of these technical advances make troubleshooting activity quicker and easier.

The advent of fluid power symbols was a giant step forward in the communication revolution which has occurred in fluid power technology. Basic fluid power texts are available. An explicit curriculum guide for fluid power programs is available; this can assist educational administrators to introduce fluid power into the educational community. One of the current guides is for the two-year Community Colleges. A second curriculum guide is structured at the technical high school level. These curriculum guides have been developed by and are available from the Fluid Power Educational Foundation, 3333 North Mayfair Road, Milwaukee, WI 53222. More advanced texts at undergraduate university design level are a continuing need as fluid power technology expands. Because fluid power covers so many and so diverse power transmission areas there is often need for significant specialization at the university level. Ultimately fluid power may be integrated into fluid mechanics text materials and courses so that the university student is exposed to fluid power science and technology. Advances in computer simulation of fluid power systems may predicate appropriate course offerings at the undergraduate engineering level, perhaps as an elective, and possibly as a mandatory subject for certain disciplines. Some courses at the university level are being made available on video format. A pioneering program at Purdue University has provided such a program for the agricultural engineer needing expertise in fluid power.

Communication channels are available between the designer and the manufacturer of hydraulically powered machinery. There is need for more and better communication channels between the machinery manufacturer, the user, and the maintenance staff. The skilled maintenance personnel cannot perform their tasks efficiently without circuit drawings, bills of materials, and properly structured maintenance manuals. Some manufacturing management experts now demand hydraulic and electrical circuit drawings encased in plastic in an accessible container *at the machine site*. In critical manufacturing operations service drawings are often mounted in a billboard format adjacent to the machine and/or material handling systems.

Cartridge valve technology is one of the answers to the second problem associated with the desirable goal of reducing the need for extensive "snake pit" piping.

Cartridge valves, whether they be of the insert, slip-in, or screw-in type, provide a virtual individual entity which is independent of the housing in which it is located. All finished dynamic surfaces are within the cartridge structure. The exterior of the cartridge is a nonmoving area which is usually sealed with an appropriate number of static elastomeric members. Sometimes an anti-extrusion ring of suitable material is also included where high shock or excessive pressures are anticipated.

Obviously there is no practical method to physically see what is happening inside any hydraulic component and especially in a cartridge valve manifold assembly on a hydraulically powered machine. The use of symbolic drawings and well-structured maintenance manuals has removed the mystery from the integrated cartridge valve manifold block as well as other hydraulic components. The quality performance, long service life, and easy maintenance of cartridge valves in manifolds, actuators, and pump housings has dictated this control system as the norm for high production or highly dependable hydraulic machinery power transmission systems. The manifold is a passive structure so that this major cost component is virtually a lifetime investment. The cartridges are easy to replace in the unlikely event of need for service.

The cartridge valve is a versatile item. It can be placed in simple housings with threaded or flanged pipe connections for prototype or short-run programs. It can be combined with other cartridge valves in useful assemblies which often become catalogued items with short delivery schedules. The cartridge valve can be installed in a machined pocket within other components such as pumps, motors, cylinders, and limited rotation actuators, or it can be encompassed within a multiple cartridge valve manifold that may have most major control functions in the one assembly. When two or more machine circuits are to be built it may be most economical to use the manifold concept. Certainly, when many machines are to be built in large quantities (e.g., typical mobile, agricultural, or similar hydraulically powered machines) then the cartridge valve concept is of prime consideration. Time for basic troubleshooting and repair are minimized.

HYDRAULIC SYMBOLS

The introduction and development of fluid power symbols through the efforts of the Joint Industry Conference (JIC) of the 1940s and subsequent activities in the American Society of Mechanical Engineers (ASME), ANSI, ISO 1219, and others must be considered a monumental step in the development of communication channels for the fluid power industry.

Graphic communications, like languages, are easiest to learn if they can be directly identified with the associated subjects. This presentation is intended to acquaint you with the basic symbols, terminology, and rules associated with the use of graphic symbology for fluid power systems. The specific use of the symbols will be expanded as you progress through this book. A working knowledge of the material in this chapter will simplify your work in any fluid power system. It is especially valuable in those systems involved with complete manifold assemblies. The symbols are the most practical and easiest route to follow the flow and power pattern through the power transmission system.

Hydraulic symbols are used throughout the world in the design, operation, and maintenance of fluid-power systems. A knowledge of hydraulic symbols will enable you to read and understand circuit diagrams and other important drawings of fluid-power systems. This knowledge is a valuable aid in your work with all types of hydraulic equipment. Some of the symbols presented in this section are called *simplified symbols*. Other, more complex symbols are explained in later chapters as they relate to specific components (valves, controls, etc.). Simplified symbols are widely used in commercial circuit diagrams. More complex symbols may be used when a basic valve or control is modified to make it perform a new or different function. Many of the symbols used for cartridge valve assemblies and other emerging technologies have been developed using the rules spelled out in the fluid power symbol systems.

CIRCUIT ELEMENTS

A *hydraulic circuit* consists of two or more fluid-power devices connected by piping or tubing. Connections within a manifold assembly may be by drilled, cast, or machined passages. In a diagram or drawing of a hydraulic circuit, a solid line (Figure 1.1a) represents a *pipe*, a *tube*, or some *other conductor* capable of handling major flow in the circuit appropriate to the rate of flow and pressure level to be contained. This type of pipe is often termed a *major conductor*; it conveys fluid to various parts of the circuit.

Pilot lines are usually of much smaller fluid-carrying capacity (and diameter) than

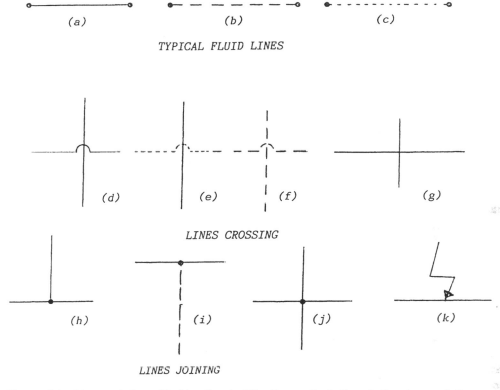

Figure 1.1 Line symbols. a, Working line. b, Pilot line. c, Drain line. d, Crossing work lines. e, Crossing work and pilot line. f, Crossing drain lines. g, Alternate crossing work lines. h, Joined work lines. i, Pilot line joining working line, j, Intersecting working lines. k, Electric line.

major lines. Figure 1.1b shows a pilot line. The length of each dash is at least 20 line-widths, with the space between dashes approximately 5 line widths. The fluid pressure in pilot lines is usually the same as that within the major piping. Pilot-line pressure may, however, be any value needed in the circuit.

Fluid-drain and *air-exhaust lines* (Figure 1.1c) may be made of lighter weight material and may not be able to withstand as high pressures as pilot or major pipes. The length of the dashed lines and the space between them in the symbol are about equal, each being approximately 5 line-widths.

Pipes that cross each other (but are not interconnected) are drawn with a small loop (Figure 1.1d). Note that Figure 1.1d shows two *major lines* (or pipes) crossing, but not interconnecting. Figure 1.1e shows a major line being crossed by a drain line (or pipe). (In fluid-power terminology the words *pipe, tubing, line,* and *conductor* are often used interchangeably to designate the structure that conveys the fluid from one device to another.) Figure 1.1f shows two pilot lines crossing each other without physical interconnection. Figure 1.1g shows two major lines crossing without the loop. This will occasionally be encountered in some industries, particularly in processing plants; this is not a preferred drawing practice for fluid power circuits.

When pipes are joined so that there is a connecting path for the fluid to each of the adjoining pipes, a *connector dot* (about 5 widths of associated symbol lines) is placed on the diagram so that it clearly shows the point of junction. This is a preferred practice for fluid power circuits. Figure 1.1h shows two major lines joined at the point designated by the dot. In an actual circuit, this is usually a pipe or tubing "tee." Figure 1.1i shows the point at which a pilot line is connected to a major line. The pilot line is usually smaller in diameter and the connection may be a tee with a reduced-diameter outlet to the pilot line. Figure 1.1j shows two major lines intersecting. The dot indicates that these lines are interconnected. A fitting in the form of a *cross* with four connections or two tees in close proximity may be used for this connection. Note that these symbols do not indicate whether the piping is fabricated from *tube and fittings* or from other types of material. Instead, the symbols show the path of the fluid through the circuit. Electric lines are shown by Figure 1.1k.

Figure 1.2a shows how a flexible connection is indicated in a circuit diagram. This symbol indicates some type of hose or similar flexible device.

Most fluid-power systems have a tank or reservoir in which to store fluid during some part of the cycle. Pipes connected to the reservoir can affect the operation of the entire hydraulic system. Because of the importance of the location of the end of the pipe (either above or below the surface of the fluid in the tank), symbols to indicate the pipe location have been devised. A U-shaped symbol (Figure 1.2b) indicates a hydraulic-fluid reservoir or tank. The U is twice as wide at the base as the height of the uprights. In Figure 1.2c the main fluid line touches the bottom of the U, indicating that the pipe terminates *below* the

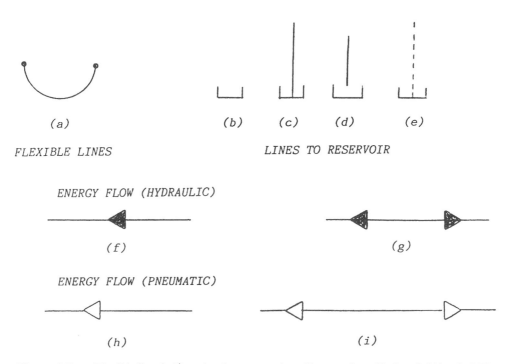

(a)

FLEXIBLE LINES

(b)   (c)   (d)   (e)

LINES TO RESERVOIR

ENERGY FLOW (HYDRAULIC)

(f)   (g)

ENERGY FLOW (PNEUMATIC)

(h)   (i)

Figure 1.2   a, Flexible line. b, Vented tank or reservoir. c, Pipe terminated below fluid level. d, Pipe terminated above fluid level. e, Drain terminated below fluid level. f, Energy flow (hydraulic) in one direction only. g, Energy flow (hydraulic) in both directions. h, Energy flow (pneumatic) in one direction only. i, Energy flow (pneumatic) in both directions.

fluid level in the tank. Drain lines from many pumps and fluid motors are connected below the fluid level so that air will not enter the device through the drain. In Figure 1.2d a major pipe is shown terminating *above* the fluid level in the tank. This arrangement is used when it is desired to drain a fluid-power machine during its idle periods. A return-to-tank port from a spring-returned cylinder may be connected to the reservoir above the fluid line so that there will be no pumping action as a result of the normal cylinder movement. Any fluid passing by the piston will go back to tank and the air passing back and forth as the cylinder reciprocates will be clean because it will be passing through the air breather in the reservoir circuit. Note that the line terminates at the top of the upright lines on the U-shaped reservoir symbol. Figure 1.2e shows a drain line that terminates below the fluid level in the reservoir.

An equilateral triangle (Figure 1.2f) is placed on a fluid line to show the direction in which the fluid is expected to travel. If a solid equilateral triangle is used, it indicates that the fluid is a liquid. An outline of an equilateral triangle (Figure 1.2h) indicates that the fluid is compressed air or a gas. If the fluid can travel in both directions at different times, two equilateral triangles facing in opposite directions are used (Figures 1.2g and i). Flow-direction triangles can be omitted from circuit diagrams unless there is a definite need for an indication of the flow direction or the fluid medium.

The symbol for a *plugged terminal* (or end) point of a pipe is indicated by an X (Figure 1.3a). Note in Figure 1.3b how the X is located in a *composite* (more than one connection) symbol. This symbol indicates that the connection to the hydraulic device is provided with a *plug* instead of a connecting line.

Nonfunctional plugs used as construction devices are not shown on hydraulic symbols and can be disregarded. Even though construction plugs may be used as an aid in cleaning a device, they are not normally shown on the circuit diagram.

A functional plug may be located at the surface of a manifold block that can house many control devices. The terminating cross of Figure 1.3c may be on a valve symbol (Figure 1.3b) or on an enclosure (*envelope*) such as that indicated by the centerline. All parts that are either within a common housing or manifolded together as a unit are indicated by the enclosure formed by the centerline (line broken by an occasional dash; Figure 1.3d). A test station may be used for a gauge connection or for connecting other test devices. Plugs may also be provided at high points in a conducting line to serve the dual purpose of bleeding air from the system on start-up and providing a gauge opening. A needle-type shutoff valve may be permanently installed in place of a plug if air must be removed frequently or if a gauge must be used for checking the pressure level when machine tooling is changed.

Figure 1.3d shows an envelope containing a *filter cartridge* and a *spring loaded check valve*, which in this installation serves as a low-pressure relief valve. The enclosing dashed line indicates that all these devices are within this common enclosure or envelope. The enclosure is usually a metal body or housing. The rotated square with the bisecting dashed line indicates the *filter*. Connecting lines are terminated at opposite corners. A wedge-shaped arrowhead with a ball between the arms indicates a *check valve*. This check-valve symbol represents a ball in a seat and is not intended as an arrow. Fluid flows from point *A* to point *B* through the filter or check valve (serving as a low-pressure relief valve) or both, depending on the state of cleanliness of the filter cartridge.

The test station on the left side of the envelope can be used to obtain a pressure reading of the fluid at that point. It may also be used to drain a series of test specimens of the hydraulic fluid before the fluid reaches the filter. At the test station, to the right of the filter, samples of the fluid can be collected after passage through the filter. This station can also be

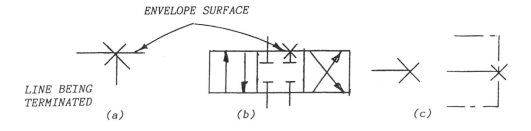

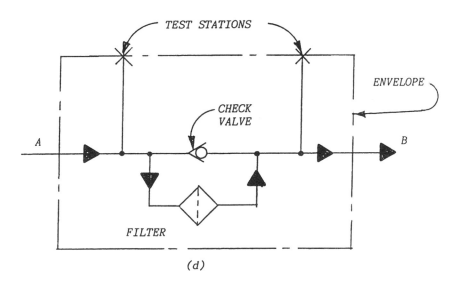

Figure 1.3   a, Plugged terminal point. b, Plugged on a four-way valve symbol. c, Test-station symbol. d, Typical test-station installation.

used for a gauge connection to check the fluid-pressure drop across the filter and check valve assembly. Test stations can be located in various parts of a fluid-power circuit to permit checking of pressure or the condition of the fluid without major piping alterations.

A fixed-size, nonadjustable *restriction within the line* is indicated by the symbol in Figure 1.4a. Dimensions of the hole in the restriction can be shown in parentheses (Figure 1.4b) for convenience in identifying the expected degree of control. Hole dimensions assist in identifying the various orifices in a complex hydraulic circuit. An idea of the rate of fluid flow and the fluid pressures to be expected in various parts of the circuit can be obtained by studying the orifice sizes. The length of the orifice and its diameter may also be given on the symbol. These dimensions are important in pressure-drop calculations.

Explanatory notes about the orifice may be desirable if the orifice is created with pins or wire in the hole to establish the desired net flow area. Use of a properly secured pin or wire may make cleaning of the orifice simpler. Changing the size of the pin or wire can provide an inexpensive method of varying the orifice area without machining the orifice

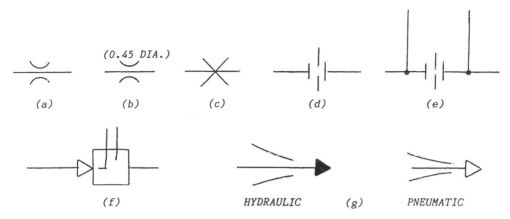

Figure 1.4 a, Fixed restriction (viscosity sensitive). b, Fixed restriction with size indicated. c, Sharp-edged orifice virtually unaffected by viscosity. d, Orifice plate. e, Orifice plate with sensing lines. f, Pitot tube. g, Nozzles.

itself. Such information is usually shown in explanatory notes. In some installations the explanatory notes may be given in a maintenance manual instead of on the circuit diagram. Ideally the information is supplied in both locations.

A sharp-edged orifice is relatively unaffected by the viscosity of the fluid (resistance to flow of the fluid). A longer flow path through the orifice will be affected by the viscosity of the fluid as it passes through the restriction. The symbols in Figures 1.4a and b anticipate a resistance to flow affected by the fluid viscosity. Figure 1.4c shows the symbol for a sharp-edged orifice which would be relatively insensitive to flow restrictions resulting from changes in viscosity. The sharp-edged orifice is widely used in compensated flow control mechanisms.

Sensing devices associated with flow rate in conductors may use orifice plates that can be changed to alter flow conditions. The symbol for an orifice plate is shown in Figures 1.4d and e. Note the sensing line before and after the orifice created by the plate inserted in the line with a fixed hole size (Figure 1.4e).

A *pitot tube* is inserted in a line to measure velocity of flow. Figure 1.4f illustrates the symbol used for a pitot tube. Hydraulic and pneumatic nozzles are indicated by the symbol shown in Figure 1.4g.

When it is desired to connect two pipes or, more often, a pipe and a hose to a hydraulic unit quickly, *quick-connect couplings* may be used. These couplings employ mechanical locks that are easily released and reassembled. Flexible hose of some type is generally used with the coupling, and tools are not needed to disconnect or reconnect. Some couplings are fitted with integral check valves that close when the coupling is disconnected. Closing of the check valve when the two halves of the quick coupling are separated prevents loss of hydraulic fluid from the unit served by the coupling. Check valves are not needed in couplings serving a hydraulic unit that is selfdraining because the fluid drains from the unit after it stops operating. Some users prefer to have the check valve plus a protective cap inserted or placed on the open end of the coupling to avoid contamination from the surrounding atmosphere.

Figure 1.5a shows two quick-disconnects coupled together without integral check

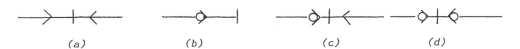

(a)   *(b)*   *(c)*   *(d)*

Figure 1.5   a, Quick-connect without check valves. b, Quick-connect with check valves (discon-nected). c, Quick-connect with one check valve (connected). d, Quick-connect with check valves in both sections.

valves. One-half of a quick-disconnect coupling with a check valve is shown in Figure 1.5b. Figure 1.5c shows a connected coupling with one half equipped with a check valve and the other half without a check valve; Figure 1.5d shows the symbol for coupled connectors with check valves in both halves. Many quick-connects used with agricultural implements and construction machinery tools have protective covers which slide into place as the units are disconnected. Some have caps attached with chain to be assembled to the open ends as the connectors are disconnected.

FLUID POWER PUMP SYMBOLS

Many variations of fluid power pump construction will be encountered when fluid power system designs are being developed. There are two basic designs that must be considered. The first is a pump that delivers a constant flow that is related to the number of revolutions of the shaft per minute. It is referred to as a constant displacement pump. The second is a pump which is capable of changing the flow rate and possibly the flow direction regardless of the rotation speed of the input drive shaft. Constant displacement pumps may have gears, vanes, pistons, screw-type units, or other geometric mechanisms that move and pressurize a specific quantity of fluid with each revolution of the input drive shaft. Variable displace-ment pumps most commonly use piston or vane assemblies. Piston assemblies may be radial, axial with a ramp type drive assembly, or bent-axis where the pistons are fitted into a rotating body that is fitted to the drive assembly with a family of universal joints to transmit the energy from the drive member to the pistons.

Centrifugal or propeller pumps are used infrequently in fluid power systems. The fluid movement resulting from the action of a centrifugal or propeller pumps is a nonposi-tive displacement function. Centrifugal and propeller pumps finds greatest use in super-charging high pressure pumps and moving fluid for test purposes, fluid conditioning, and general fluid movement activities that are only peripherally related to fluid power.

The symbols for fluid power pumps consolidate constant displacement pumps in one grouping and variable displacement pumps into another group. No effort is made to indicate finite construction factors. The details of construction are relegated to maintenance man-uals and company sales and service literature. History has shown that the choice of the symbol designers was correct in making this choice. For basic communication purposes the symbols have been kept uncomplicated. Little difficulty has been encountered with the generic characteristics of the symbol system. Several decades of use has confirmed these early decisions as being practical and efficient.

The basic envelope (or enclosure) symbol for a pump is shown in Figure 1.6a. It is a circle. Lines outside the envelope are not part of the symbol but represent pipes connected to the pump (Figure 1.6b). The direction of rotation of a shaft which drives the rotating

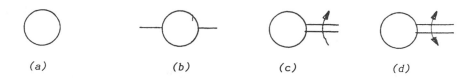

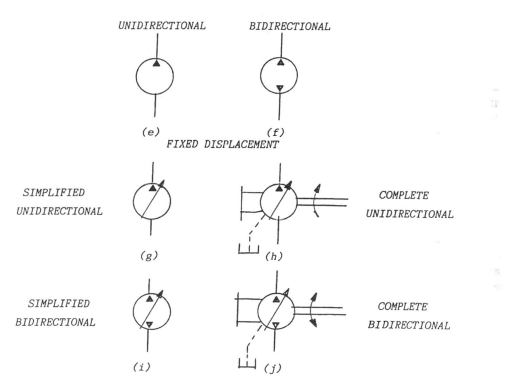

Figure 1.6 a, Basic pump symbol. b, Flow lines connected to pump. c, Shaft turns in one direction. d, Shaft turns in either direction. e, Fixed displacement, unidirectional. f, Fixed displacement, bidirectional. g, Variable displacement, simplified. h, Variable displacement, complete. i, Simplified, bidirectional. j, Complete, bidirectional.

members of a pump is shown by a curved arrow on the shaft (indicated by the two parallel lines adjacent to the circle) nearest the viewer. The curved arrow is assumed to be at the top of the shaft. Figure 1.6c shows a pump shaft that rotates in only one direction; a shaft that turns in either direction (clockwise or counterclockwise) is shown in Figure 1.6d. Direction of rotation is indicated by looking at the shaft end of the pump.

The equilateral triangle used to show energy flow through a conducting line serves a similar function in a pump symbol. A single, solid, equilateral triangle is used to show the energy-flow pattern in a fixed-displacement unidirectional pump symbol. Figure 1.6e has but one equilateral triangle which shows the direction of flow through the pump. The line opposite the equilateral triangle indicates the intake line to the pump. The discharge line from the pump adjacent to the equilateral triangle is assumed to be the supply into the immediate circuit. Bidirectional pumps that are capable of delivering pressurized liquid from either port are indicated by the symbol of Figure 1.6f.

A slash arrow across the symbol (Figure 1.6g) indicates that the pump displacement can be varied. Figure 1.6g shows the simplified symbol, while Figure 1.6h shows the complete symbol indicating manual control by the parallel lines at the left with the vertical terminating line. The drain, if included, is always shown in the complete symbol (Figures 1.6h and j). Direction of rotation is also included, if pertinent.

Bidirectional, variable-displacement pumps are illustrated by the addition of the slash arrow and the second equilateral triangle (Figure 1.6i). The complete symbol (Figure 1.6j) includes the drain and manual-control-device symbol. Many intricate pump controls will be encountered. The details of the control may be shown by symbols or by information in the maintenance manual. It is preferable to show as much information as possible in the pump symbol. Common practice is to show direction of rotation of a pump in a specific desired pattern when the pump is capable of rotation in either direction (Figure 1.6j).

SYMBOLS FOR HYDRAULIC MOTORS

A hydraulic motor is often considered an inverted pump which shares some common parts. While this may be a bit simplistic it is true that a hydraulic motor symbol needs to show that the energy provided at the input to the motor is turned into rotary force and motion. The relationship is obvious and the basic circle used for pump symbols is also used for rotary fluid motors. A solid equilateral triangle shows the direction of energy and fluid flow in the symbol, depicting a fluid motor, in much the same manner as a pump. The primary difference is that fluid-energy flow is *away* from the pump, whereas energy flow is *into* the fluid motor, where it is converted into rotary force and motion (Figures 1.7a and b).

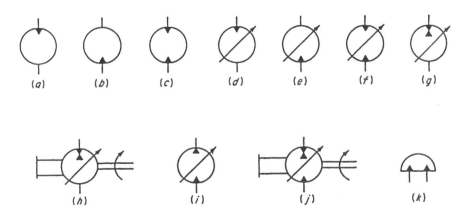

Figure 1.7   a, Basic symbol for a unidirectional motor. b, Reverse rotation of a unidirectional motor. c, Fixed displacement, bidirectional motor. d, Variable displacement, unidirectional motor. e, Reverse-rotation, variable displacement, unidirectional motor. f, Bidirectional, variable displacement motor. g, Variable motor in one direction, pump in other direction, simplified symbol. h, Complete symbol for (g). i, Variable unit functional as pump or motor in one direction only—simplified symbol. j, Complete symbol for (i). k, Limited-rotation motor.

Bidirectional operation is shown with two triangles (Figure 1.7c). Variable displacement of the rotating group is shown by the addition of the slash arrow (Figures 1.7d–f).

The variable-displacement unit of Figure 1.7g (simplified) or 1.7h (complete) functions as a pump in one direction and as a motor in the other direction. The motor, driving the cable drum of a mobile crane, often encounters this type of service. Raising a load requires the motor function. Lowering the load requires that the motor change its function and serve as a pump so that the output fluid can be restricted to provide the desired lowering speed.

The symbol of the variable unit in Figure 1.7i (simplified) or 1.7j (complete) shows a device that can function as a motor and as a pump in the same direction. For example, a mixer drive requires that energy be applied to start the unit into motion. When the mass is to be stopped, the motor becomes a pump. The delivery of the pump is then restricted to provide the needed deceleration of the mass.

SYMBOL FOR LIMITED ROTATION MOTOR OR ROTARY ACTUATOR

A motor or rotary actuator, which has a limited rotating pattern that does not permit continual rotation in one direction, is illustrated in Figure 1.7k. Many of these devices rotate less than 360 degrees prior to reversal. Devices of this type are used for tilting special railroad cars, steering the front wheel of a three-wheel tractor, rotating conveyor turnaround devices, and similar applications. Miniature units are finding use in robotic material handling devices.

LINEAR MOTORS OR HYDRAULIC CYLINDERS

Linear motors, consisting of a piston and cylinder, are represented by the symbols shown in Figure 1.8. *Diaphragms* and other linear *actuators* are represented by similar symbols. A *single-acting cylinder* is shown in Figure 1.8a. Fluid enters the left end of the cylinder and forces the piston to the right. An external force, such as gravity acting on moving machine members, produces reverse movement of the piston and forces the fluid back into the hydraulic circuit. Figure 1.8d shows a modification of the single-acting cylinder fitted with an *internal spring* to return the piston to its original position. The spring space or *cavity* has an external drain that terminates *above* the reservoir fluid level. Were this drain terminated below the fluid level, the piston might act as a pump during the reverse stroke and fill with fluid from the reservoir.

A *double-acting cylinder* with a single rod end is shown in Figure 1.8b. Sometimes it is desirable to use both ends of the piston rod. To achieve this, the piston rod is extended to pass through both ends of the cylinder (Figure 1.8c). This arrangement is a *double-acting, double-rod cylinder*.

Integral devices to decelerate the piston and its associated load at the ends of the stroke are illustrated by the rectangle shown attached to the piston in Figure 1.8e. The addition of the slash arrow (Figure 1.8f) shows that the cushion or deceleration device is adjustable.

If the relationship of rod diameter to cylinder bore is significant, it may be desirable to indicate this (Figure 1.8g).

Telescopic cylinders, as the name suggests, provide long strokes by the use of multiple sections that are nested within the concentric assembly as they retract. Figure 1.8h

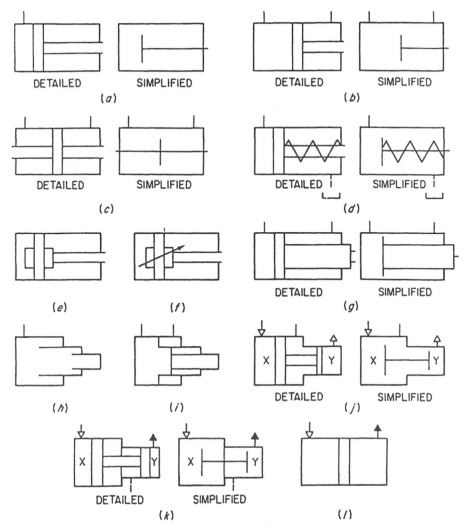

Figure 1.8   a, Single-acting cylinder, diaphragm, or other linear actuator. b, Double-acting, single-rod cylinder. c, Double-acting, double-rod-end cylinder. d, Double-acting cylinder, pressure in one direction, spring return, single-rod cylinder with drain above liquid level. e, Cylinder assembly with fixed cushion, advance and retract. f, Adjustable cushion, advance and retract. g, Oversize rod. h, Telescoping cylinder, single-acting. i, Telescoping cylinder, double-acting. j, Pressure intensifier or volume amplifier, pneumatic. k, Hydraulic and pneumatic intensifier. l, Air-oil transformer.

shows a single-acting assembly and Figure 1.8i shows a double-acting telescopic cylinder. Combinations of two basic cylinders with differently sized pistons are used to provide hydraulic pressure intensification or, inversely, volume amplification. A single, common rod may be employed. The symbols shown in Figures 1.8j, 1.8k, and 1.8l are not intended to show construction. They indicate a pressure-intensification function or, under inverse conditions, a volume amplification.

The assembly in Figure 1.8j shows that a pneumatic pressure x is transformed into a higher pneumatic pressure y. Figure 1.8k shows that a pneumatic pressure x is transferred into a higher *hydraulic* pressure y.

Figure 1.8l shows an air-oil actuator. Equipment of this design transforms a pneumatic pressure into a substantially equal hydraulic pressure (or vice versa). Such equipment is often used to obtain a smooth flow by the addition of a hydraulic-flow control mechanism. Typical units are available for power feeds on drill presses, etc.

SYMBOLS FOR HYDRAULIC VALVES

The basic symbol for a hydraulic valve is a rectangle, termed the *valve envelope*. As with pumps and motors, the valve envelope represents the valve enclosure or body. Lines within the envelope show the flow directions between the valve inlet and outlet openings, termed *ports*. To show the change in flow conditions when the valve is opened or closed (*actuated*), two systems of symbols are used: 1) single envelope and 2) multiple envelopes.

Single-envelope symbols are used where only one flow path exists through the valve. Flow lines within the envelope indicate static conditions when the actuating signal is not applied. The arrow can be visualized as moving to show how the flow conditions and pressure are controlled as the valve is actuated.

Multiple-envelope symbols are used when more than one flow path exists through the valve. The envelopes can be visualized as being moved to show how flow paths change when the *valving element* in an envelope is shifted to its various positions.

Figure 1.9a shows a typical single envelope; multiple envelopes are shown in Figure 1.9b. Flow lines to the valve ports are added to the two types of basic valve envelopes in Figure 1.9c and 1.9d. Note in Figure 1.9d that the ports are shown at the center envelope segment.

A variety of valve-port conditions are shown in Figure 1.10. Thus, the double arrow in Figure 1.10a denotes a blocked or closed port in a single envelope where flow may be expected to pass in either direction when the valve is actuated. The blocked condition is indicated by the fact that the arrows do not line up with the lines coming to the single square. Figure 1.10b shows the same condition in a multiple-envelope valve. The single envelope signifies that an infinite number of positions may be encountered between the closed (normal) and open position, such as might be expected with a relief valve where the opening is a function of the orifice size that must be created to provide the needed pressure level in an automatic function.

The multiple envelope of Figures 1.10b and c indicates a finite number of positions which the directing elements can be normally expected to assume. The normal position of Figure 1.10b is usually indicated by the left block. The normal position of the unit (Figure

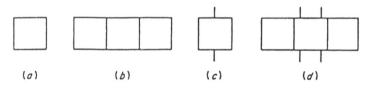

Figure 1.9 a, Basic envelope. b, Multiple basic envelopes. c, Ports attached to basic envelope. d, Ports attached to normal multiple basic envelopes.

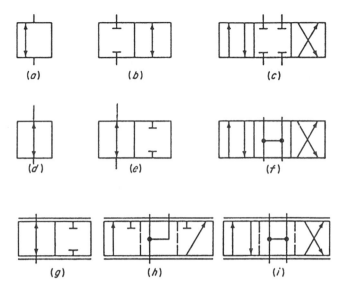

Figure 1.10   a, Blocked port. b, Blocked port, finite positions. c, Three-position, blocked ports in neutral, finite positions. d, Open port, infinite positions. e, Normally open, finite positions. f, Open center, three finite positions. g, Two-position, infinite crossover positions. h, Two-position, infinite crossover position, three-way diversion valve. i, Two-position, infinite crossover position, four-way valve with neutral fully open.

1.10c) is the neutral position in which all the ports are blocked. This condition is referred to as a *closed center*. Shifted to conditions in the left block, the flow is in the direction indicated by the arrows. When shifted to the right, the flows are reversed. This is referred to as a *four-way function*; flow at one input port can be directed to separate outlets or blocked, while a return line is directed to an opposite port in a predetermined pattern.

A through flow is indicated by the symbol of Figure 1.10d. An infinite number of positions of the flow-control elements from full-open to closed can be anticipated with this design. The two-element symbol of Figure 1.10e indicates two finite positions. The three-element symbol of Figure 1.10f indicates three finite positions, with the neutral (rest) position indicating the full interconnection of all associated lines.

The two-element symbol of Figure 1.10g is provided with parallel lines at the top and bottom. This signifies that it represents a valve with two finite positions; however, the nature of the application is such that it may be shifted to infinite positions in the process of moving from one position to the other. An example is a cam-operated two-way valve used to decelerate flow; the cam may follow a random pattern in the functional movement from open to closed position.

The diversion valve of Figure 1.10h indicates that all ports are interconnected as the flow-directing element is moved. The dashed lines between the blocks indicate that the neutral position is not a finite one. The information contained in the center block indicates only the conditions as the valve elements are shifted from one finite position to the other. The open-center condition of the symbol shown in Figure 1.10i would not normally be shown in a two-block symbol. The addition of the third position between the dashed lines provides the needed data without indicating three finite positions.

Figures 1.11a and b summarizes the basic symbol, normal, and actuated conditions for normally open and normally closed single-envelope valves. Figure 1.11c shows an alternate arrow design which indicates that one port is always connected to one flow path, whether normal or actuated.

The valve in Figure 1.12a is a *spring-loaded relief valve*, piped so that the discharge flows to a reservoir having the outlet below the fluid level. The signal source for a relief valve is ahead of the normally closed element. When the pressure rises to a predetermined value, the valve opens and the flow passes through the valve, and it can be imagined that the arrow is aligned as in Figure 1.12a, where it is shown in the actuated position. A sequence valve (Figure 1.12b) has a similar function to that of the relief valve. Because the outlet port will be directed to a pressurized area, it is necessary to provide an external drain for the spring pocket or for draining a pilot mechanism if it is included in the design.

An unloading valve usually drains internally because the secondary port normally connects to the major return line. If pressure is expected on the valve outlet (as might be encountered in especially long return lines to tank), an external drain may be necessary; the operating signal (Figure 1.12c) is usually from an external source.

Two pressure-reducing valve symbols (Figures 1.13a and b) show the basic symbol as a two-way valve that is normally open and that may assume infinite positions during normal function. The *signal source* is at the valve outlet or *low-pressure secondary port*. The spring pocket is normally externally drained back to the tank with a minimum restriction. Figure 1.13b shows the addition of a major port discharging to the tank. The flow directing element indicates potential flow in either direction. Normal functional flow is directly through the

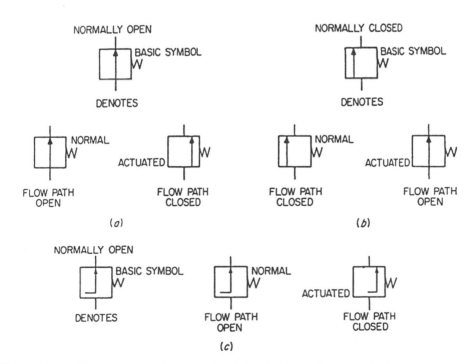

Figure 1.11   a, Normally open, single-envelope valve. b, Normally closed, single-envelope valve. c, Alternate arrow design to indicate that one port is always connected, whether actuated or at rest.

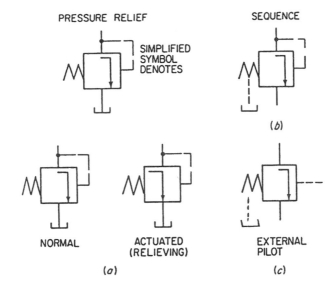

Figure 1.12 a, Relief valve. b, Sequence valve. c, Unloading valve.

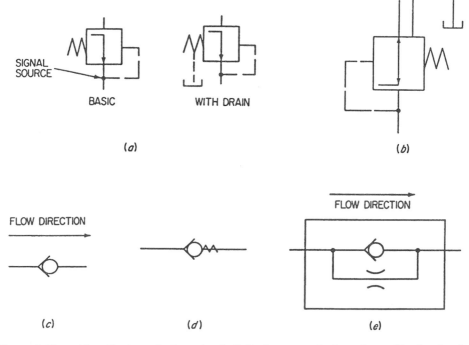

Figure 1.13 a, Nonrelieving reducing valve. b, Relieving-type reducing valve. c, Check valve, free. Opens if the inlet pressure is higher than the outlet pressure. d, Spring loaded. Opens if the inlet pressure is greater than the outlet pressure plus the spring pressure. e, Unit allowing free flow in one direction but restricted flow in the other.

valve. As a predetermined reduced-pressure value is attained, flow diminishes to a balance point. If external forces cause secondary pressures to increase, the control element functions beyond the usual levels and diverts excess fluid to the reservoir to ensure an appropriate maximum pressure level.

The symbol for a *check valve* is shown in Figure 1.13c; flow in this valve is from left to right. Do not confuse this symbol with an arrowhead. The circle represents the moving element of the check valve, often a *ball*, and the slant lines represent the *seat*, against which the ball is pressed when the valve shuts.

Figure 1.13d shows the check valve with a bias spring. The valve opens if the inlet pressure is greater than the outlet pressure plus the spring pressure.

In Figure 1.13e the check valve is fitted with an orifice. The box symbol provides an indication of the composite nature of the device and shows the fluid-energy flow path. The fluid flows freely from left to right through the check valve. Should the check valve shut and the fluid flow reverse itself, the orifice would restrict the amount of fluid passing from right to left.

Two *pilot-operated check-valve* symbols are shown in Figures 1.14a and b. This type of valve may have an internal piston mechanism that may either force the valve open or

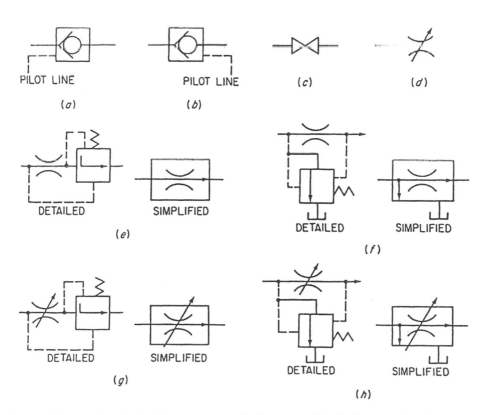

Figure 1.14   a, Pilot check valve, pressure applied to open valve. b, Pressure relaxed to open. c, Manual shutoff valve. d, Variable-flow-control valve. e, Pressure-compensated flow-control valve with fixed orifice. f, (e) with fixed output and relief port to reservoir. g, (e) with variable orifice. h, (e) with variable output and relief port to reservoir.

hold it closed, depending on the design. The dashed line in Figure 1.14a represents the pilot line used to conduct fluid to the piston mechanism to hold the check valve open. A drain line may be shown leading from the lower part of the enclosure. A baffle device is designed for certain models. The cavity between the baffle and upper surface of the operating piston must be drained. Generally, this drain is terminated above the fluid level in the reservoir. In some circuit configurations the drain from this barrier area in the pilot check may be pressurized during one portion of the circuit to retract the actuating piston assembly rapidly and permit the check to close quickly or move the actuating piston out of the direct fluid flow. In Figure 1.14b the pilot line is arranged to hold the valve closed. When pilot pressure is *relaxed* (relieved) in this valve, fluid may pass in either direction through the valve in certain designs. In other designs the check will perform in a conventional manner until pilot pressure is applied. At that time the check is held closed so that flow is not possible in either direction. Releasing the pilot pressure then permits the usual function.

A manual shutoff valve of either the *gate*, *ball*, or *globe* type is represented by the symbol shown in Figure 1.14c. This type of valve is not intended for control of rate of flow in an infinite adjustment pattern.

For several types of *variable-flow-rate-control valves*, a symbol similar to that for an orifice is used. A slash arrow is drawn across the symbol (Figure 1.14d), indicating that flow can be varied by changing the orifice size. Pressure compensation is indicated by the symbols in Figures 1.14e–h. Variations in inlet pressure do not affect the rate of flow.

During the development of the symbols for compensated flow control mechanisms a symbol was used similar to the simplified symbol of Figure 1.14e with the addition of a vertical arrow within the box. This vertical arrow was to indicate a compensating mechanism. To designate a temperature compensation mechanism a vertical line with a bulb at the bottom was used to indicate a thermometer bulb (indicating temperature considerations). These symbols are widely used but are not accepted as official.

MULTIPLE-ENVELOPE VALVE SYMBOLS

In this type of valve the basic symbol consists of: 1) an envelope for each operating position of the valve, 2) internal flow paths for each valve position, 3) arrows indicating flow direction through the valve, and 4) external ports at the normal or neutral valve position.

Figure 1.15a shows a *two-position, three-connection valve* in its neutral or rest position. The external ports are always shown at the neutral or rest position in the valve symbol. Actuated, this valve is shown in Figure 1.15b. Note that the port is shown as connected to the flow path represented by the arrowhead in the right-hand box. A four-port

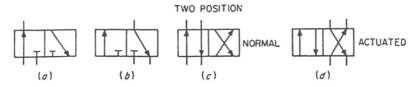

Figure 1.15   a, Normal position of a two-position, three-connection valve. b, Actuated position of a two-position, four-connection valve. c, Normal position of a two-position, four-connection valve. d, Actuated position of a two-position, four-connection valve.

two-position valve is shown in Figure 1.15c in the rest position. Actuated the ports will connect as shown in Figure 1.15d. Note that the ports are shown as connected to the flow path represented by the arrowhead in the right-hand box.

Three-position, four-connection valves have three working positions. Figure 1.16a shows a three-position valve in the normal rest position where there is no need for showing potential transitional conditions. In this neutral position all ports are blocked. In Figure 1.16b the valve is in its actuated position with fluid shown flowing through the right-hand box. Flow is shown through the left-hand box in Figure 1.16c. An additional two positions may be shown as in (Figure 1.16d) with dashed dividers, indicating that the inner squares show a transitional condition. Thus, the symbol in Figure 1.16d indicates that all valve ports are interconnected as the valve is shifted from neutral to an extreme position or from the completely shifted position back to neutral. In the neutral position of the valve of Figure 1.16d, two ports are shown as being connected and two ports are blocked. Typical neutral conditions for variety of three-position, four-connection valves are shown in Figure 1.17. Identifying letters for the various ports may vary with different manufacturers. The existing standards do not cover the identification of the ports other than the U-shaped symbol indicating return to the tank and the equilateral triangles indicating energy flow.

A complete symbol for a three-position, four-connection, spring-centered, solenoid-controlled, four-way, directional-valve with three ports interconnected and one port blocked in neutral is shown in Figure 1.18.

Figure 1.19a shows a two-position, three-connection, spring-offset, solenoid-controlled valve. When the solenoid is deenergized, the spring moves the internal working mechanism to a position where there is a direct flow through the valve, as indicated by the box adjacent to the spring symbol. This is considered the normal position, and the external lines would be connected to the left-hand box. The two-position, four-connection, solenoid-controlled valve in Figure 1.19b does not have a rest position (no motion position) estab-

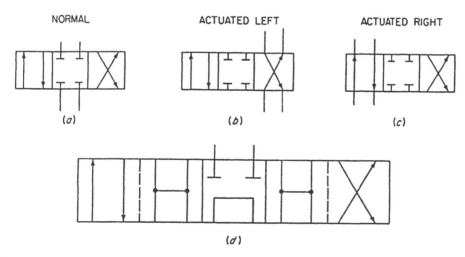

Figure 1.16 a, Normal position of a three-position, four-connection, four-way, directional-control valve. b, Actuated position of valve in (a); envelope to left. c, Actuated position of valve in (a); envelope to right. d, Dash dividing line to indicate transitory conditions as valve is shifted from neutral to extreme positions or from extreme shifted positions back to neutral.

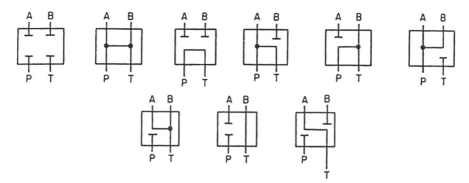

Figure 1.17 Four-way, directional-control valve neutral configurations. Identifying letters vary with manufacturers.

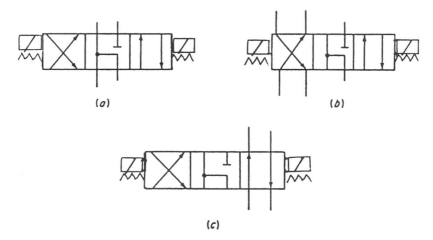

Figure 1.18 Complete symbol for four-connection, three-position, spring-centered, solenoid-control, directional-control valve; three ports interconnected and one port blocked in neutral. The three positions of the valve are: a, Neutral, centered by springs when solenoids are deenergized. b, Actuated, envelopes to right by energizing left solenoid (adjacent to block, showing energy flow). c, Actuated, envelopes to left by energizing right solenoid.

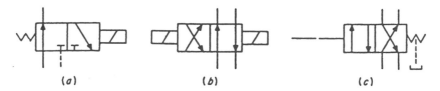

Figure 1.19 Directional-control valves. a, Two-position, three-connection, spring-offset, solenoid-control valve. b, Two-position, four-connection, solenoid-control valve. c, Two-position, four-connection, spring-offset, pilot-operated valve.

lished by an energy source such as a spring or gravity. Because of this, the external connecting lines are usually shown connected to the box in what might be considered the rest position in the circuit in which the valve is used. The right-hand solenoid would be energized prior to the valve coming to a rest position; the external lines would then be connected to the right-hand box (Figure 1.19b).

Pilot-operated, two-position, four-connection, spring-offset valves usually have some provision to drain the end cap containing the spring mechanism. Figure 1.19c shows the drain, with the pilot connection on the opposite end. External lines are shown as being connected to the box adjacent to the spring which would determine the rest position when a signal is not present.

Figure 1.20a shows a three-position, four-connection, spring-offset, mechanically controlled valve with one external port plugged. In this symbol the center or neutral position is illustrated, even though the unit does not stay in the neutral position for any fixed length of time. The symbol informs the viewer that all ports will be blocked on the neutral transitional crossover point until the reversal function is started. The dashed line between blocks indicates that the valve does not stay in neutral because the bias spring will urge the flow-directing elements to a predetermined position. The knob on the operator attached to the left end of the symbol in Figure 1.20a indicates a manual lever. Five connections to a three-position valve can be made (Figure 1.20b). The drain connections prevent full pressure being exerted on the shaft seals. At each end of Figure 1.20c the two parallel lines with the vertical bars at the ends indicate a mechanical operator, which might be a stem, lever, cam, or knob. A multiplicity of connections can be incorporated within a two- or three-position valve. For example, the two-position, eight-connection, mechanical-control valve in Figure 1.20c can provide a predetermined flow pattern with a variety of external connections.

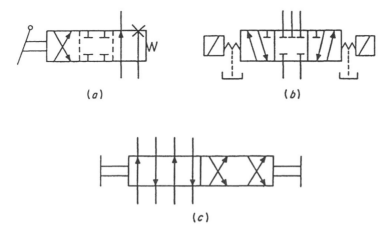

Figure 1.20 Directional-control valves. a, Three-position, four-connection, spring-offset, mechanical-control valve; one external port plugged. b, Three-position, five-connection, spring-centered, solenoid-control valve. c, Two-position, eight-connection, mechanical-control valve.

SYMBOLS FOR TYPES OF CONTROL

When a spring is used in a control function, it is indicated by a wavy zigzag line (Figures 1.21a and b). It is placed so that one end touches the envelope. Pilot pressure may emanate from an internal or external source. Figure 1.21c shows the hyphenated line coming from an external source. A pilot function can result from either an application or a release of pilot pressure. Figure 1.21d shows the equilateral triangle pointing away from the major symbol, indicating that the operation is by release of pressure. Pilot actuation by differential pressure is shown in Figure 1.21e.

There are general symbols for manual operation (Figure 1.21f), push button (Figure 1.21g), lever (Figure 1.21h), pedal or treadle (Figure 1.21i), and plunger or tracer (Figure 1.21j). A detent symbol (Figure 1.21k) is most generally used with mechanically or pilot-operated valves. Some small-size pilot valves with dual-solenoid operation employ light detents to prevent spool drift because of machine vibration. The detent mechanically holds the flow-directing mechanism in the desired position until a specific force value is applied to cause movement from that position to another finite position. Individual solenoids (Figure 1.21l) or a dual solenoid, with two windings operating in opposite directions (Figure 1.21m) or in a reversing motor (Figure 1.21n), provide electrical signal application. Combinations of actuators are common. Applications may dictate use of solenoid *OR* pilot or solenoid *AND* pilot actuation (Figure 1.21o). AND, OR, AND/OR symbols are shown in Figures 1.21p and q. Electrohydraulic servo valves are shown in Figures 1.21r–t.

RESERVOIR SYMBOLS

There are two classes of hydraulic-system reservoirs: *vented reservoirs* or tanks are indicated by a U-shaped symbol (Figure 1.22a) and *pressurized tanks* (Figure 1.22b). A line from a manifold may be vented to permit free flow of fluid back to the reservoir (Figure 1.22c).

MISCELLANEOUS SYMBOLS

Figure 1.22d shows the symbol for a gauge for pressure or vacuum. Figure 1.22e shows the symbol for a temperature-indicating gauge. The sensing bulb may be shown at a remote point if it assists in clarifying the circuit details. Flow meters (Figure 1.22f) are of two types. *Rate* is indicated by the left symbol. The symbol at the right indicates a *totalizing meter* such as those used to indicate water consumption in homes. In industry a totalizing meter may be inserted in a pump line with an electrical readout mechanism to determine when a pump should be rotated in a multipump circuit.

Rotating joints in fluid conducting lines are indicated by the symbol in Figure 1.22g when only one line is within the assembly. Two or more lines are shown (Figure 1.22h), which also indicates a drain from the seal assembly.

In the pressure switch, fluid energy is translated to an electrical signal. At a predetermined pressure value the mechanism converts fluid pressure as it enters the valve into motion that actuates an electrical switch. The symbol for the pressure switch shown in Figure 1.22i shows the contacts involved.

The top accumulator symbol in Figure 1.22j is used worldwide as a single symbol to designate any type of accumulator. ANSI graphical symbols also use the symbol of Figure 1.22j as a general-purpose symbol; however, they also recommend use of more explicit symbols such as those shown in Figure 1.22j for spring-loaded, gas-charged, and weighted accumulators. The symbol for a spring-loaded accumulator encloses a zigzag and parting line to indicate use of a spring as an energy-storage medium. The symbol for a pressurized-gas accumulator (hollow equilateral triangle plus dividing line) usually indicates an inert gas, such as nitrogen, as the energy-storage medium. A small rectangle on the dividing line is used to indicate a weighted-type accumulator. The symbol for a receiver in Figure 1.22j is usually associated with compressed air. It may also be used to show an auxiliary tank connected to a gas-charged accumulator to provide a larger quantity of gas under pressure to increase the energy-storage capacity.

Symbols for fluid conditioners employ a square resting on a corner with the associated major lines connected at the corners on a horizontal plane (Figure 1.22k).

The symbol for a heater (Figure 1.22l) shows the introduction of energy into the major line. The cooler symbol (Figure 1.22m) indicates heat dissipation or energy flowing from the major line. Dual arrows (Figure 1.22n) indicates that energy can flow in either direction. A heater, cooler, or temperature-controller symbol can be provided with external lines, indicating whether transfer is by means of gas or liquid. A solid equilateral triangle in a line, terminating at the midpoint on the diagonal of a square, indicates the medium is a liquid; a hollow equilateral triangle indicates a gas or compressed-air or a radiator-type heat exchanger, such as that used on conventional, internal-combustion, water-cooled engines.

The symbol for a filter or strainer employs a dashed line bisecting the square (Figure 1.22o). No attempt is made to differentiate between the various types of filters or strainers. The position of the symbol in the circuit will usually provide a clue as to the type. Strainers are found most often on pump-suction lines, where introduction of major, large-size contaminants into the pump is prevented by the strainer. Major users of fluid power systems have outlawed strainers on pump suction lines that cannot be serviced without opening the main reservoir. Strainers buried in a reservoir rarely get serviced and a pump could sustain greater damage from a restricted strainer than what might occur with other contaminants. Suction filters are available which are designed with protective gauge or indicator device to alert the operator when the filter is saturated to a point which could affect the operation of the pump. Obviously, a clean tank with return line and/or other continuous cleaning mechanisms for the operating fluid leads to troublefree operation of the system.

Filters may be found in any part of a circuit. However, machine tool circuits often employ high-pressure, in-line filters just ahead of critical components. Mobile equipment may use return-line filters most effectively because of the expected motion of the tank and resulting agitation of the fluid. There are no fixed rules as to strainers and filter usage, so the material list on a drawing may be needed to indicate the specific information. History has proven that a clean tank with well-conditioned fluid and properly positioned and serviced filters lead to long, trouble-free service from the hydraulic power transmission system.

Much pneumatic equipment is used in association with hydraulic systems. Because of this, it is usual to find pneumatic conditioning device symbols on hydraulic drawings. Pneumatic directional-control valves, used in circuitry, employ symbols similar to the equivalent hydraulic function. Hollow, equilateral, energy-flow triangles provide the clue as to the fluid medium. Figure 1.22p shows a separator used to remove moisture from pneumatic lines. The symbol of Figure 1.22q illustrates a combination filter and separator for pneumatic lines. A desiccator, or chemical drier, is shown in Figure 1.22r.

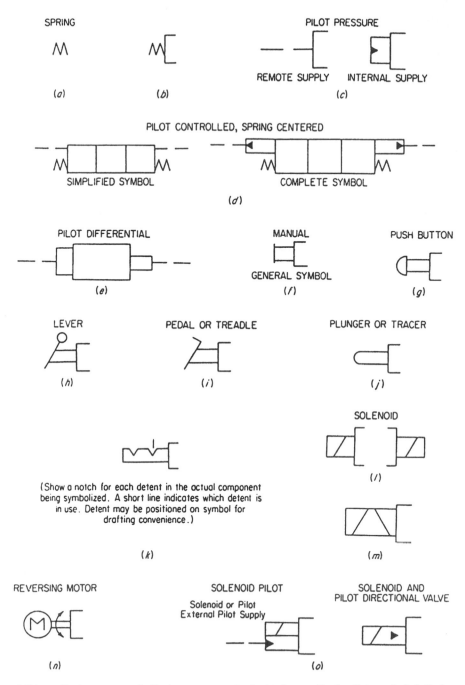

Figure 1.21   a, Spring actuator. b, Spring actuator attached to box. c, Basic pilot symbol. d, Spring-centered pilot assembly. e, Pilot differential. In the symbol the larger rectangle represents the larger control area (i.e., the priority phase). f, Manual operator used as general symbol without indication of specific type (i.e., foot, hand, leg, arm). g, Push button. h, Lever. i, Pedal or treadle. j, Plunger or tracer. k, Detent. l, Solenoid. m, Solenoid with two windings operating in opposite directions. n, Reversing motor. o, Solenoid OR pilot, solenoid AND pilot directional valve. p, Composite actuators—AND,

COMPOSITE ACTUATORS (AND, OR, AND/OR)

BASIC One signal only causes the device
to operate

OR One signal OR the other signal causes
the device to operate

AND One signal AND a second signal both
cause the device to operate

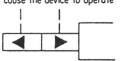

AND/OR The solenoid AND the pilot OR
the manual override alone causes the
device to operate

(*p*)

The solenoid AND the pilot OR the manual
override AND the pilot

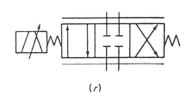

The solenoid AND the pilot OR a manual override
AND the pilot OR a manual override alone

(*q*)

SINGLE-STAGE
ELECTROHYDRAULIC SERVO VALVE

(*r*)

TWO-STAGE WITH MECHANICAL FEEDBACK

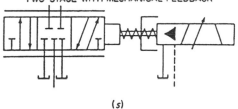

(*s*)

TWO-STAGE WITH HYDRAULIC FEEDBACK

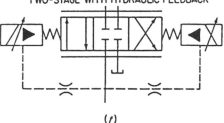

(*t*)

OR, AND/OR. q, Pilot, manual and solenoid combination actuators. r, Electrohydraulic servo valve: a unit which accepts an analog electrical signal and provides a similar analog fluid-power output. Single stage. s, Two-stage electrohydraulic servo valve with mechanical feedback, with indirect pilot operation. t, Two-stage with hydraulic feedback, with indirect pilot operation.

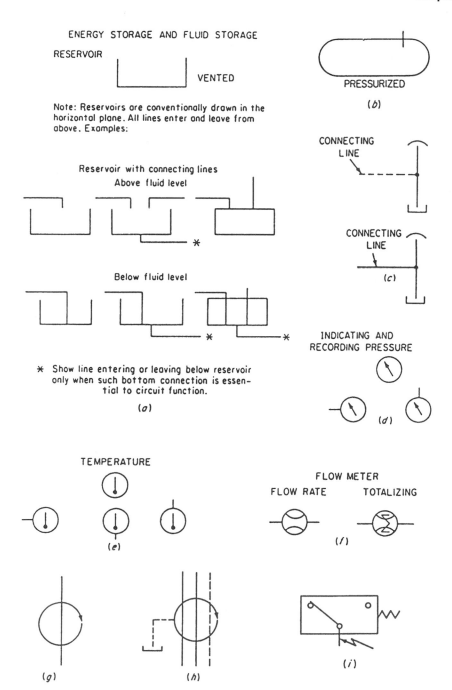

Figure 1.22 a, Vented reservoir. b, Pressurized reservoir. c, Vented manifold. d, Pressure or vacuum gauge. e, Temperature gauge. f, Flow, flow rate, and totalizing. g, Rotating connections with one flow path. h, Rotating connection with two major working lines, pilot line and drain. i, Pressure switch. j, Accumulators and air receiver. k, Fluid conditioner. l, Heater. m, Cooler. n, Temperature controller. o, Filter strainer. p, Separator. q, Filter separator. r, Desiccator. s, Air-line lubricator. t, Electric motor. u, Heat engine.

ACCUMULATORS

Spring loaded

Gas charged

Weighted

RECEIVER, for air or other gases

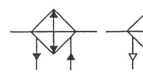

(*j*)

COOLER

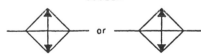

Inside triangles indicate heat dissipation

Corners may be filled in to represent triangles

(*m*)

FLUID CONDITIONERS

Devices which control the physical characteristics of the fluid

(*k*)

HEAT EXCHANGER HEATER

Inside triangles indicate the introduction of heat

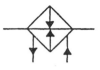

Outside triangles show the heating medium is liquid

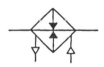

Outside triangles show the heating medium is gaseous

(*l*)

TEMPERATURE CONTROLLER

The temperature is to be maintained between
two predetermined limits

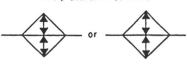

(*n*)

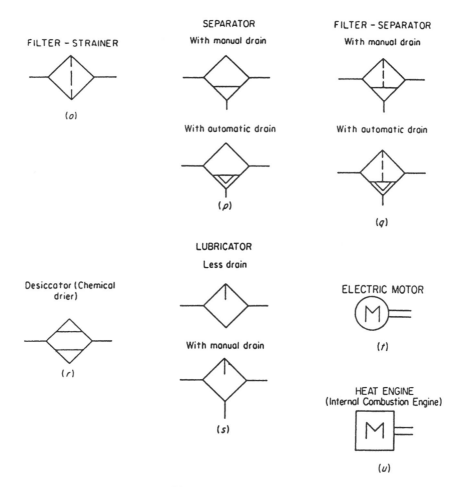

Figure 1.22   Continued

Pneumatic lines usually require the introduction of a lubricant into the line to minimize rust and corrosion. The oil may also serve to assist in lubricating operating devices in the circuit, such as air-piloted hydraulic valves. Figure 1.22s shows the units with and without a manual drain.

Figure 1.22t shows the symbol for an electric-drive motor. The symbol for Figure 1.22u illustrates a heat engine, such as a steam, diesel, or gasoline engine.

COMPOSITE SYMBOLS

A *component enclosure* (Figure 1.23a) may surround a composite symbol or a group of symbols to represent an *assembly*. It is used to convey more information about component connections and functions. The enclosure indicates the limits of the component or assembly. External ports are assumed to be on the enclosure line and indicate connections to the components. Flow lines cross the enclosure line without loops or dots.

In a typical component enclosure, one dash approximately 20 line-widths long is

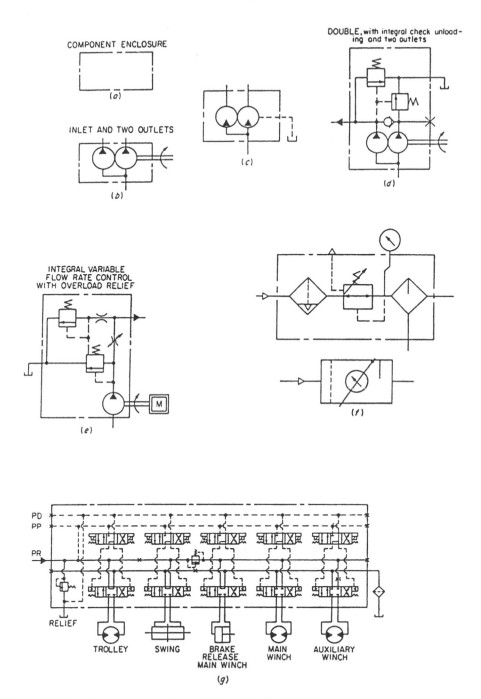

Figure 1.23 a, Component enclosure. b, Double fixed-displacement pump with one inlet and separate outlets. c, Double fixed-displacement motor with one inlet and separate outlets. (This assembly is used as a flow divider/flow combiner.) d, Double pump with integral check valve, unloading and relief valves. e, Single pump with integral variable-flow-rate control and overload relief valve. f, Control assembly for air supply to machine circuit. (Upper figure complete symbol, lower figure simplified symbol.) g, Mobile stack valve assembly.

separated from the continuing line by a space of approximately five line-widths. The solid connecting line running lengthwise or around the corner should be approximately 10 dash-lengths long. This can be varied but should include at least one dash to each side, top, and bottom. This enclosure is often referred to as a *centerline-type enclosure.*

An enclosure containing two pumps on a common shaft with a common suction line and independent discharge lines is shown in Figure 1.23b. An enclosure containing two motors of equal size with a common shaft used to equalize or split the flow of fluid can be shown in a similar manner (Figure 1.23c). The drain may be included in some designs if the motor case is not designed for pressure equal to the working pressure.

Composite symbols, such as those in Figure 1.23d, can be easily interpreted if a logical study pattern is followed. The best way to do this is to trace each line from its source to its endpoint. To start, follow the suction line from the tank to the pump. Starting from the pumps opposite the rotating shaft (left pump), note that there is a working line passing through the enclosure to the circuit. Immediately beyond the pump discharge is a line connected to a check valve. The position of the check valve indicates that the flow from this pump can never pass through the check valve. Thus, this point can be considered a dead end as far as the delivery of the pump is concerned. Directly above the left pump is the symbol for a relief valve. The discharge of this relief valve is piped through the enclosure to the tank. Below the relief valve is a pilot line going to a valve in the right-hand pump circuit. This pilot line usually provides only a signal, and movement of fluid in it is relatively small. The pilot line will also be a dead end when the signal to this valve is completed. The left-hand pump, opposite the drive shaft, is provided with a circuit that can normally supply all machine pressure requirements, but it may not provide sufficient flow for the required speed. The extra pump is included in most circuits of this type to provide additional speed through added volume of fluid.

The discharge of the right-hand pump is connected to the input of a normally closed, two-way valve. Between the pump discharge and the inlet of the two-way valve a connection is provided to the inlet side of a check valve. This check valve permits discharge of the right-hand pump to connect with the delivery of the left-hand pump, combining the volume of both pumps for delivery to the circuit. When the pressure signal to the left-hand pump becomes high enough, it causes the pilot-operated two-way valve to open, permitting the right-hand pump discharge to pass freely through this valve back to tank. This return flow combines with the discharge from the relief valve and passes through the enclosure to the tank.

The symbol in Figure 1.23d does not necessarily indicate precise values of the fluid pressure in the various parts of the circuit, but certain relationships can be assumed. The highest fluid pressure will be developed by the left-hand pump. The piloted two-way valve must be adjusted to open at a pressure *less* than the relief-valve opening pressure. If this adjustment is incorrect and the pressure required to open the piloted two-way valve exceeds that of the relief valve, then the delivery of both pumps will combine and pass through the relief valve to the tank when the circuit is not accepting fluid under pressure. The testing station shown provides a place to install a gauge so that the point at which the right-hand pump starts to divert fluid to the reservoir can be checked and adjusted if necessary.

By tracing each line in this way and by knowing the basic symbols used, a complete understanding of all functional operations of a hydraulic circuit can be obtained. Figure 1.23e shows a flow-and-relief circuit enclosure. Trace its various parts in the same way as described above.

Figure 1.23f illustrates the composite and simplified symbols for filters, regulators, and lubricators in an assembly that serves to clean, lubricate, and regulate air diverted from a central system to be integrated into a fluid-*power* circuit.

Stackable valves, such as those used widely in mobile hydraulic circuits, may include many functional units in one manifolded assembly. Note the components included in the circuit of Figure 1.23g. The first two directional-control valves employed for major flow directing are in series; the last three are in a parallel arrangement. All pilot valves are in a parallel supply and return to the tank circuit. Filtering is accomplished within the return to the tank line.

International standards have proven to provide best communications. However, five types of hydraulic diagrams have been used in commercial practice. These are: 1) graphical diagrams, 2) circuit diagrams, 3) cutaway diagrams, 4) pictorial diagrams, and 5) combinations diagrams.

Graphical diagrams (Figure 1.24) show each piece of hydraulic apparatus, including all interconnecting pipes, by means of approved standard symbols. Diagrams of this type provide a simplified method of showing the function and control of each component.

Graphical diagrams are incorporated into specific *circuit diagrams*. Many large manufacturing groups have special forms designed to show special features that are useful to their designers and maintenance people. Several examples are included in the standards issued by the American National Standards Institute.

Cutaway diagrams require significant drawing time to produce the needed detailed

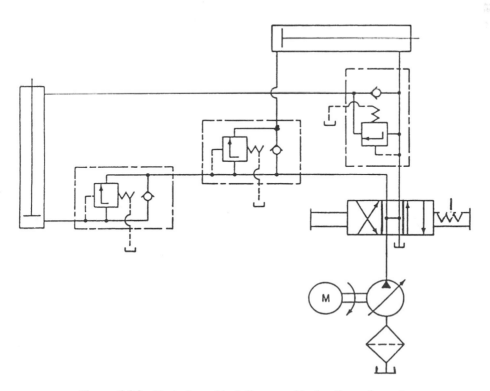

Figure 1.24 Typical graphical diagram of hydraulic equipment.

information. Generally, this type of diagram is used by manufacturers of fluid-power components to illustrate features within their product structure.

Pictorial diagrams are rarely used in state-of-the-art fluid-power communications. Prior to the development of the ISO International Standards symbols and circuit diagrams, the pictorial diagram with much verbiage served as the communications medium.

Combination diagrams, like the pictorial diagram, have become obsolete and can be found most often in old machine tool instruction books.

SUMMARY

Graphical symbols for fluid-power circuits are international in nature. They provide clear-cut circuit information, regardless of language barriers. In a well-prepared circuit diagram, every part of a hydraulic circuit must be clearly shown, with a minimum need for auxiliary notes. Symbols are available for most commercial components. By following basic rules, it is possible to create easily understood symbols for special applications.

Hydraulic symbols show connections, flow paths, and the function of the components represented.

Symbols can indicate conditions that occur during transition from one flow-path arrangement to another. Symbols can also show if a device is capable of infinite positions or if it contains only finite working positions. A direct relationship is indicated between operator devices and the resulting flow paths. Symbols do not usually indicate construction or circuit values, such as pressure, flow rate, and temperature. Symbols do not indicate the location of ports, the direction of spool movement, or the position of control elements in actual components. (Modified symbols may show some of these features in special units.)

Hydraulic symbols may be rotated or reversed without altering their meaning except in the case of vented lines and lines to the reservoir, where the symbol must indicate whether the pipe terminates above or below the fluid level. Line-width does not alter the meaning of symbols: A symbol may be drawn any suitable size. The symbol size may be varied for emphasis or clarity.

Where flow lines cross, a loop is used except within a symbol envelope. The loop may be used in this case if clarity is improved.

In multiple-envelope symbols, the flow condition, shown nearest a control symbol, takes place when that control is caused or permitted to actuate. Each symbol is drawn to show the normal or neutral condition of a component unless multiple circuit diagrams are furnished showing various phases of circuit operation. Arrows are used within symbol envelopes to show the direction of flow through the component in the application represented. Double-ended arrows indicate reversing flow.

External ports are located where flow lines connect to the basic symbol, except where a component enclosure is used. External ports are at the intersection of flow lines and the component enclosure symbol, where the enclosure is used.

SAMPLE MACHINE

To relate the troubleshooting procedures described in this book to an actual machine, we have selected a surface grinder in which the machine's reciprocating motion passes a workpiece back and forth under a grinding wheel. Proper operating requirements include:

- accurate table inching for setup
- variable table speed
- smooth table reversal and intermediate stopping
- control of grinding speed when travel direction is reversed
- grinding motion in forward direction that is controllable to adjustable fixed positions

Knowing the surface grinder's performance specifications is a prerequisite to successful troubleshooting. The machine malfunction procedure that is discussed in Chapter 2 provides a logical sequence of thought and action which can be applied to the surface grinder or to any other hydraulically powered machine that develops a malfunction.

2
Identification

Identify the machine malfunction and the components that might be the cause.

The troubleshooting process must follow a logical sequence. This rule cannot be over-emphasized. The step-by-step chart and flow diagram of Figure 2.1 provide an orderly plan to determine the cause or causes of a machine's malfunction.

INITIAL TROUBLESHOOTING PROCEDURE

Step 1. Any fluid power actuator is operating improperly if you notice:

- no movement
- movement in the wrong direction
- erratic movement
- incorrect speed
- creep
- incorrect sequence
- incorrect force

Whatever the reason, the problem should be defined in terms of *flow*, *pressure*, or *direction*.

Step 2. Using the circuit diagram, identify each component in the system and determine its function.

Step 3. Compile a list of components that could possibly cause the problem (Figure 2.2). Thus, slow actuator speed can be related to *flow* even though this may, in turn, be due to lack of pressure. Remember that a leaking or improperly adjusted relief valve (i.e., a *pressure* control valve) could affect flow to the actuator.

Step 4. Arrange the list of components in rough order of priority based on your past experience and the ease with which they can be checked.

Step 5. Conduct a preliminary check of each component on the list. Look at installation, adjustment, signals, etc. Also determine if any component shows abnormal symptoms such as excessive temperature, noise, vibration, etc.

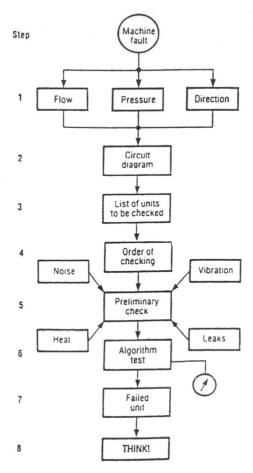

Figure 2.1 Eight basic steps to thoughtful, organized troubleshooting. Courtesy of Vickers, Incorporated.

Figure 2.2 Compile a list of components that could possibly cause the problem. Courtesy of Commercial Intertech.

Step 6. If the preliminary check reveals no obviously faulty component, conduct a more exhaustive repeat test of each using additional instrumentation—*without removing* any unit from the system. (Algorithms detailing these important tests will be provided in later chapters.)

Step 7. The instrument checks should identify the faulty component. Someone then must decide whether to repair or replace.

Step 8. Before restarting the corrected machine, stop and consider the cause and *consequence* of the failure. For example, if the failure was caused by contaminated or over-heated fluid, you should anticipate additional component failures until you take remedial action against that underlying cause. If a pump failed mechanically and debris entered the hydraulic system, the system should be thoroughly cleaned before a new pump is installed and operated.

Again, *think* about what caused the failure and its potential consequences.

3

Beginning the Troubleshooting Process

Use your senses to begin the troubleshooting process

Unfortunately, many companies neglect or ignore the internal record-keeping that is necessary to perform a logical troubleshooting sequence. It may seem foolish to review machine basics on equipment that has run for years, but in many cases, personal observation is the only way to determine operating parameters that can lead you to where problems originate.

The preliminary checklist (Figure 3.1) can help you find problem sources that have nothing to do with the machine's hydraulic system.

On the other hand, hydraulic systems may show signs of leakage or excessive heat, noise, or vibration without any adverse effect on machine performance—at least not at first. But these early signals should not be ignored. Continue your preliminary check by using the human senses of sight, touch, and hearing to identify and pinpoint the problem area.

It is important to remember that a machine may exhibit abnormal symptoms because of a problem that might actually be present elsewhere in the system. For instance, pump cavitation or aeration might be taking place in a remotely located hydraulic power station. In such a case, the fault/cause/remedy checklist (Figure 3.2) should be followed to identify the problem source.

If a preliminary check does not find the problem, a more exhaustive examination, similar to the procedure suggested in the machine fault checklist must be conducted (Figure 2.1).

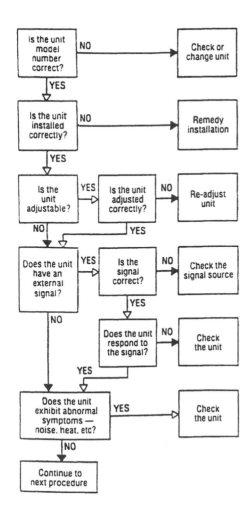

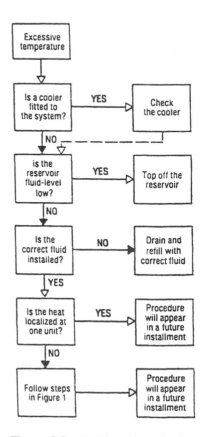

Figure 3.1 Preliminary checklist includes some obvious but essential questions. Courtesy of Vickers, Incorporated.

Figure 3.2 Fault/cause/remedy checklist for excessive temperature. Courtesy of Vickers, Incorporated.

4
Fault, Cause, and Remedy

Noise, heat, vibration, and leakage can interfere with the proper operation of a hydraulic system or provide warning of future problems. Maintenance and troubleshooting personnel can help avoid these problems, simply by using their natural senses of sight, touch, and hearing in an organized pattern.

As previously mentioned, a hydraulic system may malfunction or operate improperly because of one or more problems located away from the system. One example would be cavitation or aeration in a pump located in a remote pump house. Under such conditions it might help to use the fault, cause, remedy (FCR) review process outlined in Figure 4.1.

Seldom is an *entire* system at fault even though several malfunctions may occur in close sequence. In a large system, first examine that part of the hydraulic installation where the faulty component is mounted. Follow these steps:

1. Determine which of the four basic problem symptoms seem to apply: excessive heat, noise, vibration, or leakage. If the problem is heat, see Chapter 3. Vibration and leakage investigation will be covered in later chapters.
2. If excessive noise appears to be the problem, consult the chart in Figure 4.2. It provides a selection process that considers many common conditions associated with excessive pump noise.
3. If, after this examination, you conclude that noise is caused by cavitation, check the system following the suggested FCR sequence shown in Figure 4.3.

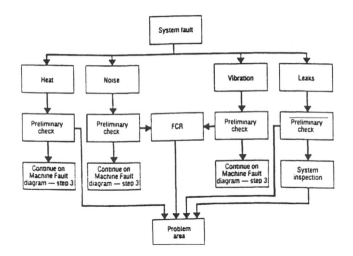

Figure 4.1 System fault analysis. Courtesy of Vickers, Incorporated.

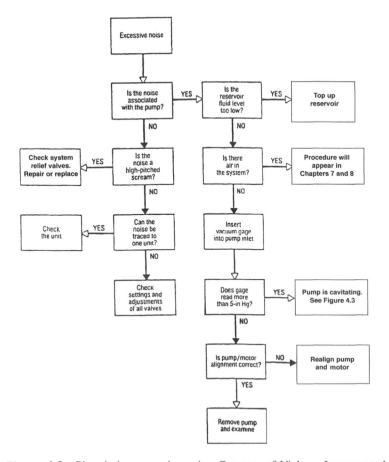

Figure 4.2 Pinpointing excessive noise. Courtesy of Vickers, Incorporated.

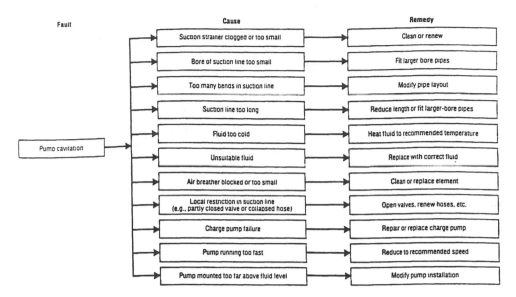

Figure 4.3 Pump cavitation causes and remedies. Courtesy of Vickers, Incorporated.

5
Vibration

Although vibration is a mechanical defect, it significantly affects hydraulic system performance.

It is always important to remember that a *hydraulic* system consists of many *mechanical* components. Mechanical components are subject to mechanical problems. Vibration is a case in point.

Technically, excessive vibration is a mechanical, not a hydraulic problem. In a well-designed hydraulic system, vibration is neither identified with nor related to the shifting action of hydraulic controls, the linear movement of cylinders, the rotation of pumps and motors, or the oil itself. Yet, mechanical imbalance—which results in vibration—can and often is the cause of hydraulic problems that maintenance persons must troubleshoot and deal with. Figures 5.1 and 5.2 are detailed troubleshooting procedures for excessive vibration and fluid aeration, respectively.

LONG-TERM CONSEQUENCES

The trouble is that engineers often tend to design to resolve an immediate problem, but frequently fail to consider the longer-term consequences of their designs. One example of lack of design foresight is *component inaccessibility* on many machines. Replacing a pump and achieving the required alignment of electric motor and hydraulic pump—to avoid vibration—in a machine is often difficult and sometimes nearly impossible because electric motors and pumps are often buried inside the machine. The initial assembly is made without problems, but after other machine elements are in place, maintenance, repair, and replacement becomes a headache.

The designer should have anticipated a potential long-term maintenance problem, and not just designed for an easy, short-term assembly solution. A Band-aid fix may be a relatively cheap short-term cure but can become very expensive in the long run. *Don't keep it a secret.*

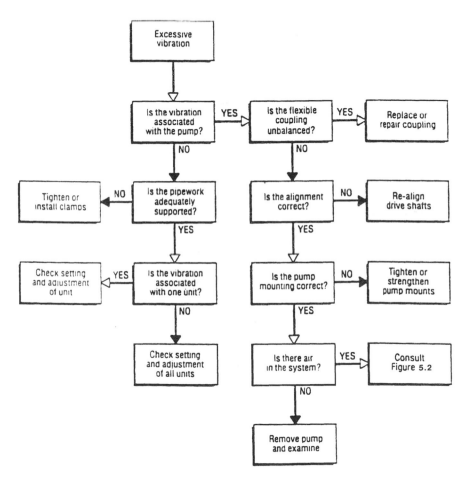

Figure 5.1 Pinpointing causes of vibration problems. Courtesy of Vickers, Incorporated.

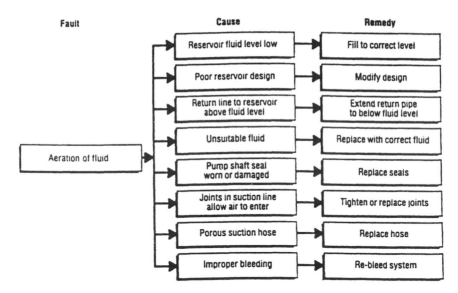

Figure 5.2 Fluid aeration causes and remedies. Courtesy of Vickers, Incorporated.

Maintenance personnel should be encouraged to think about causes and prevention of equipment problems, and to communicate their thoughts, observations, and ideas to their company's management. It is management's responsibility to review and evaluate suggestions submitted by the maintenance department and to consider reported design shortcomings when planning future equipment purchases. Otherwise, problems are just perpetuated.

6
Leakage

Eliminating leakage should be a cooperative project between a company's management and its maintenance department.

There are two types of leakage: external and internal. Because internal leakage (i.e., inside components) is a complex topic all its own, this discussion is limited to troubleshooting external leakage. Monitoring external leakage is fairly easy (Figure 6.1); stopping it is not difficult, but preventing its recurrence may be a challenge that may require piping and/or circuit changes.

Every industry has been plagued by piping system leakage. This condition exists at all levels of industry between the drawing board and/or computer aided design and computer aided manufacturing (CAD-CAM) and the scrap dealer:

- At the original equipment manufacturers (OEM) level—the delivery of new units is frequently delayed because of fluid system leakage.
- At the dealer level—sales are lost when oil is observed dripping from the unit onto the showroom floor.
- At the user level—costly work stoppages occur to replace blown O-rings and retighten fittings.

The basic causes of fluid system leakage are:

- human error
- lack of quality control
- poor protection of components in handling
- difficult if not impossible to reach fitting connections
- lack of education
- poor selection of materials
- improper design of piping or pipe routing

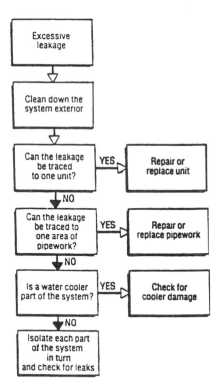

Figure 6.1 Troubleshooting procedures for hydraulic leakage problems. Courtesy of Vickers, Incorporated.

FIND THE LEAK

1. Pinpoint the leak location; make sure that the leak is not up higher and draining down:
    a. wash down leakage area.
    b. watch for the leak to show.
    c. place a paper towel above the connection, it will catch any fluid dropping from above.
2. Determine if the leak is at a valve spool, rod packing, motor or pump shaft, cracked casting, piping joint, or hose or tube. *Seepers or weepers can be hard to locate.*

TAPERED PIPE THREAD

All other factors notwithstanding, the greatest contributor to external leakage has been the tapered pipe thread. When components with pipe threads are assembled, the threads leave a space shaped like a conical helix. This helix is a ready-made leakage path (Figure 6.2). Various methods have been used to eliminate this leakage path—from mechanically crushing the threads to using sealing tapes or other sealant materials—but results are not always satisfactory, particularly when connections must be opened and remade. A re-

IN STANDARD PIPE THREADS,
THE FLANKS COME IN CONTACT
FIRST.

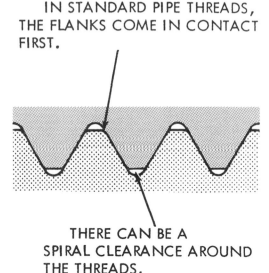

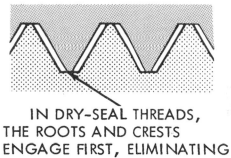

THERE CAN BE A
SPIRAL CLEARANCE AROUND
THE THREADS.

IN DRY-SEAL THREADS,
THE ROOTS AND CRESTS
ENGAGE FIRST, ELIMINATING
SPIRAL CLEARANCE.

Figure 6.2 The troublesome NPT thread.

Figure 6.3 The dry seal thread did not cure all leakage problems and, in fact, created some new problems.

designed taper pipe thread called the dry seal eliminated the helix and crushed the thread when engaged (Figure 6.3). Manufacturers found that when cast iron components with dry seal threads were tested by the manufacturer prior to shipping the dry seal thread form in the relatively soft body was damaged and new components often leaked when installed on the machine. A taper pipe thread when over-tightened may split the casting or cause distortion which causes valve malfunctions. The solution is *not to use NPT or NPTF threads at all*. Substitute straight thread and/or flange connections that are leak free.

SOCIETY OF AUTOMOTIVE ENGINEERS (SAE) STANDARD J514

The straight thread ports and fitting were developed by a consortium of fitting manufacturers, working under the umbrella of the SEA Committee on Tubing, Piping, Hoses, Lubrication and Fittings. Their joint effort, in the early 1950s, culminated in the design of a new port and fitting configuration (Figure 6.4).

The basic design concept was surprisingly simple: instead of relying on metal interference and crushing of threads to effect a hydraulic seal, as in the case with any tapered pipe threads, the committee came up with a *straight thread* design in which an elastomeric seal, such as an O-ring, does the sealing.

SAE straight thread fittings can be divided into two general groups: positionable (Figure 6.5) and nonpositionable (Figure 6.4). Nonpositionable fittings retain the O-ring with a shoulder machined on the body of the connector. Positionable fitting can be machined in many different configurations with a lock nut and backup washer (Figure 6.6).

Two alternate designs for a port entry plug to be used with the SAE straight thread

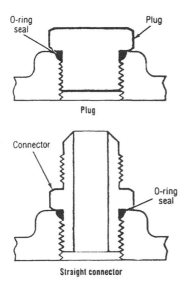

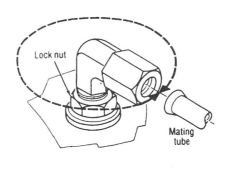

Figure 6.4 Plugs and straight connectors require no positioning.

Figure 6.5 Infinite positioning is possible with use of the backup washer and lock nut.

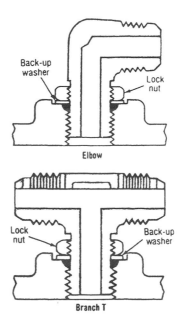

Figure 6.6 The positioning of various fitting configurations can be accomplished with the use of the backup washer and lock nut.

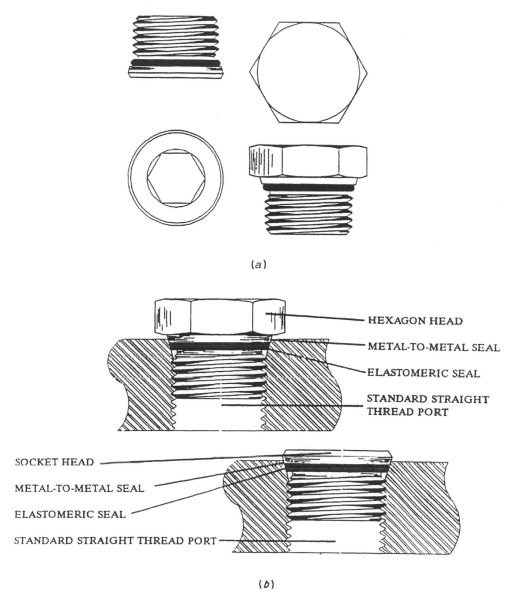

(a)

HEXAGON HEAD

METAL-TO-METAL SEAL

ELASTOMERIC SEAL

STANDARD STRAIGHT
THREAD PORT

SOCKET HEAD

METAL-TO-METAL SEAL

ELASTOMERIC SEAL

STANDARD STRAIGHT THREAD PORT

(b)

Figure 6.7   Port entry plugs with metal-to-metal seal to back up the elastomeric seal for use with SAE and ISO 6149 straight thread ports (NWD International, Inc.; Patent No. 4934742).(a) Optional hexagon socket and hexagon head design. (b) Cut-away view of typical installation showing metal-to-metal seal backing up the elastomeric seal. Courtesy of NWD International, Inc.

port are illustrated in Figure 6.7 (NWD International, Inc.; patented). The tapered metal-to-metal seal supports and retains the elastomeric seal.

Figure 6.8 illustrates an alternate design for a positionable fitting that employs the SAE J514 standard straight thread O-ring boss port. The metallic intrusion in the taper near the surface of the boss provides high integrity metal-to-metal seal, even without the O-ring.

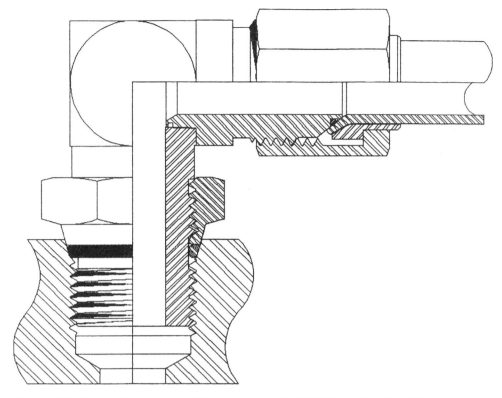

Figure 6.8 Adjustable port entry fitting designed to fit SAE J514 straight thread O-ring boss port (NWD International, Inc.; Patent No. 4934742). Courtesy of NWD International, Inc.

This fitting design also provides an elastomeric seal between the inner metal surface of the flared tubing and the matching nose of the fitting (NWD International, Inc.; patented). Figure 6.9 illustrates the tube just prior to assembly. Figure 6.8 illustrates the unit assembled with the elastomer providing the fluid barrier in conjunction with the two metal-to-metal sealing surfaces.

SAE Straight Thread O-Ring Seal

Causes of leakage:
1.   Elbows loosen up after short service.
2.   O-ring leakage after short service.
3.   O-ring leakage after long service.
4.   Instant leakage upon start up.

Causes may be either human error or faulty parts!

Cures:
1.   Replace O-ring seals and start over.
     a. Jam nut and washer must be to the back side of the smooth portion of the elbow adapter.

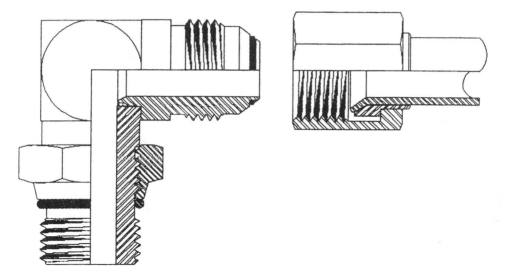

Figure 6.9 Elastomeric sealing ring inserted between the inner surface of the tube flare (37 degree or 45 degree) and the hose of the fitting providing both metal-to-metal and elastomeric sealing facilities (NWD International, Inc.; Patent No. 4458926). Courtesy of NWD International, Inc.

      b. It is very important to lubricate the O-ring.

      c. Thread into port until washer bottoms onto spot face. (Make certain that the spot face is large enough for the washer or the hex of the straight adapter.

      d. Position elbows by backing up the adapter.

      e. Tighten jam nut.

  2. Lubrication is *essential* for the O-ring.

      a. Fitting engaged to point where O-ring touches face of boss (Figure 6.10). Lubrication on O-ring permits it to move in direction D.

      b. When O-ring and boss are dry, rotary motion of assembly can cause friction and O-ring can move in direction C.

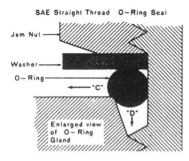

Figure 6.10 A lubricated seal will move into the pocket as intended in the D direction. An unlubricated seal may move in the C direction.

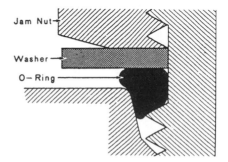

Figure 6.11 An unlubricated seal moving in the C direction will be pinched as the jam nut is tightened.

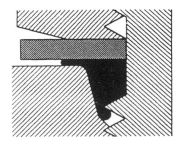

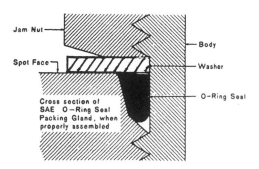

Figure 6.12 The pinched elastomer will cold flow leaving the pocket unfilled and ultimately loose.

Figure 6.13 A properly positioned seal will be retained by the washer and jam nut providing the desired seal at the interface between the conductor and the component.

    c. Jam nut and washer cannot bottom fully if the O-ring is between the washer and the face of the boss (Figure 6.11).

    d. The compressed elastomer between the washer and the boss face will cold flow out from compression and the fitting will be loose and usually leak (Figure 6.12).

    e. A properly assembled fitting will result in the washer securely containing the elastomeric seal with the face of the jam nut parallel with the washer and spot face (Figure 6.13).

This design depends on compression or squeeze of the O-ring, the jam nut, washer, face of boss, and the body threads to hold the elbow tightly in position.

    3. If some elbows loosen up and others do not:

      a. If the jam nut loosens, the whole fitting becomes loose.

      b. SAE permits a chamfer on both sides of the jam nut (Figure 6.14).

      c. When both sides of nut are chamfered, there is much less compression area to squeeze onto the washer when tightened (Figure 6.15).

      d. When the washer side of the jam nut is left flat, the contact and squeeze area is greatly increased and makes a big difference (Figure 6.16).

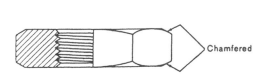

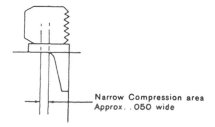

Figure 6.14 Jam nuts may be chamfered on both faces.

Figure 6.15 A chamfer at the washer side will reduce compression area.

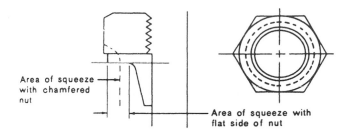

Figure 6.16   The area of squeeze with a flat on the jam nut at the interface with the washer and spot face will increase the area of squeeze for the elastomer and overall holding ability.

If the problem persists consider a jam nut that is flat on the washer side with the jam nuts and retaining washers prepositioned to the far side of the smooth area to help eliminate human error.

The sectional view of a typical mobile-type directional valve shows the straight thread, pocket for the O-ring, and spot face in the work port (Figure 6.17). The plug used to secure the load hold check also uses the straight thread and O-ring configuration. The supply line to the valve and the return to tank may be fitted with a four bolt flanged connection, typically in sizes of one-inch or larger.

SAE 37 Degree Flare Fitting

Causes of leakage. Most of the leaks on this connection are due to the lack of tightening or human error. You cannot tell if the nut has been tightened by just looking at the connection. If it is more than finger tight, you cannot tell from observation how tight it is (Figure 6.18).

Torque wrenches are good only when they are used. You must rely on the user to be sure they get used on all joints. The user must depend on his memory to know if he or she has tightened all of the joints.

Cures. Here is a foolproof method of tightening. Anyone can tell if the joint was tightened and how much:

1. Tighten the nut finger tight until it bottoms the seats.
2. Mark a line lengthwise on the nut and extend it onto the adapter. Use an ink pen or marker.
3. Using a wrench, rotate the nut to tighten. Turn the nut the amount shown on the chart of Figure 6.19.

Troubleshooting a joint that has been tightened properly. Remove the line and inspect for:

1. Foreign particles in the joint (wash them off).
2. Cracked seats (replace them).
3. Seat mismatched or not concentric with the threads (replace the adapter).
4. Deep nicks in the seat (replace faulty parts).
5. Excessive seat impression, indicating a material that is too soft for high pressures and that threads will stretch under high pressure (replace the part).
6. Phosphate treatment, an etching process which if overdone leaves a rough sandpaper-like surface (replace faulty parts).

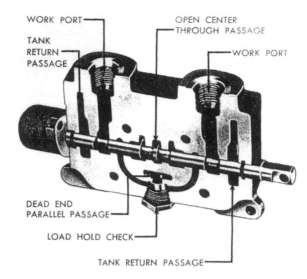

Figure 6.17 Cross-section of a typical mobile-type directional-control valve illustrating the work port with straight thread, O-ring pocket, and spot face. Courtesy of Commercial Intertech.

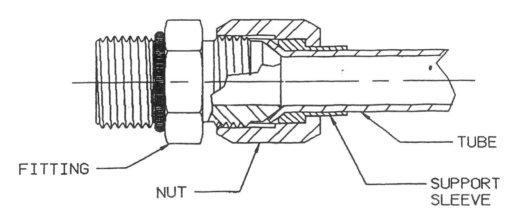

Figure 6.18 Sectional view of 37 degree flare (ISO8434-2).

Line Size	Rotate No. of Hex Flats
— 4	2½
— 5	2½
— 6	2
— 8	2
—10	1½ - 2
—12	1
—16	¾ - 1
—20	¾ - 1
—24	½ - ¾

Note: *The misalignment of the marks will show how much the nut was tightened and, best of all, that it has been tightened.*

Figure 6.19 Suggested rotation of nut to insure proper tightness.

7. Chatter or tool mark—high and low spots on seats (replace faulty parts).
8. Inspect for SAE 45 degree nuts which may not provide a tight joint (use all SAE 37 degree flare parts).

Many of the leakage problems on this type of connection will not show until the unit has had a few hours of service. All items except (1) are quality control problems that are usually found on marginal parts.

O-RING ON THE NOSE OF THE FITTING (FLARE-O®)

An additional degree of leakage control is offered by machining a groove in the face of the angle in the 37 degree fitting (Figure 6.20). The metal-to-metal seal is maintained for mechanical strength on either side of the O-ring and the captive elastomer can serve as a resilient static seal (NWD International, Inc.; patented).

FLARELESS FITTING

Flareless fittings (Figure 6.21) can be assembled and removed from the circuit many times without loss of sealing ability. Many different types of flareless fittings have been produced in the past. Flareless fittings must also be tightened to the values recommended by the manufacturer of the fittings.

O-RING FACE SEAL FITTING

Brazing the tube to a machined sleeve provides a secure and leak tight joint. Assembling the sleeve and tube to a straight thread adapter with a static O-ring also provides a leak proof assembly (Figure 6.22). Proposed metric O-ring face seal fittings are shown in Figure 6.23.

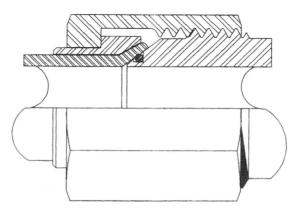

Figure 6.20 FLARE-O fitting with resilient O-ring between the nose of the fitting and the tube surface (NWD International, Inc.; Patent No. 4458926). Courtesy of NWD International, Inc.

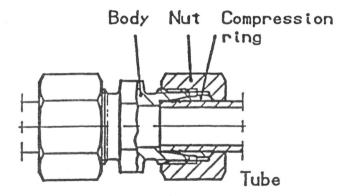

Figure 6.21 24 degree cone flareless fitting (ISO8434-1).

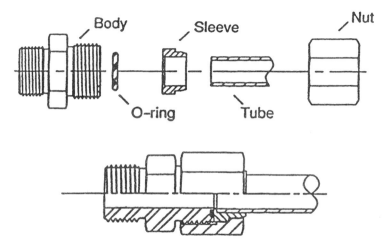

Figure 6.22 O-ring face seal (Proposed ISO8434-3).

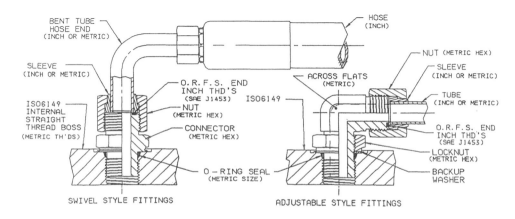

Figure 6.23 Proposed metric O-ring face seal.

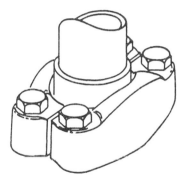

Figure 6.24 SAE J518c 4-bolt split flange connection.

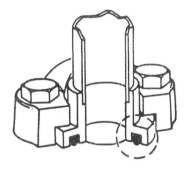

Figure 6.25 The face seal and its retainer must be parallel with the component face.

SAE (J518c) SPLIT FLANGE CONNECTION

The SAE 4-bolt split flange connection is a face seal (Figure 6.24). The shoulder containing the seal must fit squarely against the mating surface and be held there with even tension on all bolts (Figure 6.25).

The shoulder, in which the O-ring is nested, protrudes past the flange halves by 0.010–0.030 in. This is to insure that the shoulder will make contact with the mating accessory surface before the flange does. The flange halves overhang the shoulder on the ends so that the bolts will clear the shoulder.

Potential for leaks. This connection is very sensitive to human error and bolt torquing. Because of the shoulder protrusion and the flange overhang, the flanges tend to tip up when the bolts are tightened on one end, in a seesaw fashion. This pulls the opposite end of the flange away from the shoulder and when hydraulic pressure is applied to the line, it pushes the shoulder back into a cocking position (Figure 6.26).

Corrective measure. All bolts must be installed and torqued evenly. Finger tightening with the use of feeler gauges will help to get the flanges and shoulder started squarely. When the full torque is applied to the bolts, the flanges may bend down until they bottom on the accessory. This also causes the bolts to bend outward (Figure 6.27). Bending of the

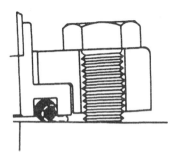

Figure 6.26 All bolts must be installed and torqued evenly to the proper value.

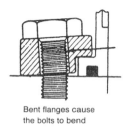

Bent flanges cause the bolts to bend

Figure 6.27 Excess torque will cause the flange to bend down until they bottom on the accessory.

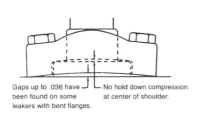

Gaps up to .036 have ⌐ No hold down compression
been found on some   at center of shoulder.
leakers with bent flanges.

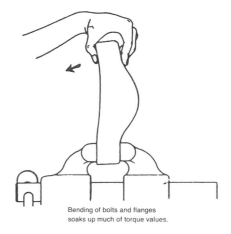

Bending of bolts and flanges
soaks up much of torque values.

Figure 6.28 Bending of the flanges and bolts tends to lift the flange off the shoulder in the center area between the long spacing bolts.

Figure 6.29 Bending of bolts and flanges soaks up much of the torque value.

flanges and bolts tends to lift the flange off the shoulder in the center area between the long spacing of the bolts (Figure 6.28).

When pipes and/or hose are joined together with this connection, the conditions become more severe because the spacing between mating flanges now is doubled and becomes a 0.020–0.060 in. gap. All conditions are now multiplied by 100%.

Bending of bolts and flanges soaks up much of torque value (Figure 6.29). High torque is required on all bolts which must be Grade 5 or better because much of the torque is lost in overcoming the bending of the flanges and bolts (Figure 6.30).

Proper installation. Lubricate the O-ring before assembly. All mating surfaces must be clean. All bolts must be evenly torqued. Do not tighten any one bolt fully before going to the next one. Torque to the values shown in Figure 6.31.

Because of the tolerance build up in all component parts plus the bolt bending, the flange halves can move sideways in direction A and B (Figure 6.32). This can lessen the shoulder contact with the flange to zero in the center area between the long bolt spacing. When flanges have a large radius on the edge "D," the leakage problem becomes even

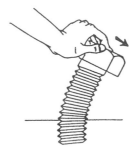

Figure 6.30 Class 5 or better bolts are needed to minimize bending.

RECOMMENDED TORQUE VALUES (USE GRADE 5 BOLTS)

Connection Size	Torque Foot Pounds
– 8	21
–12	40
–16	40
–20	60
–24	90
–32	90
–40	90
–48	175

Note: *Air wrenches tend to cause flange tipping.*

Figure 6.31 Recommended torque value for proper installation of SAE J518 4-bolt split flanges.

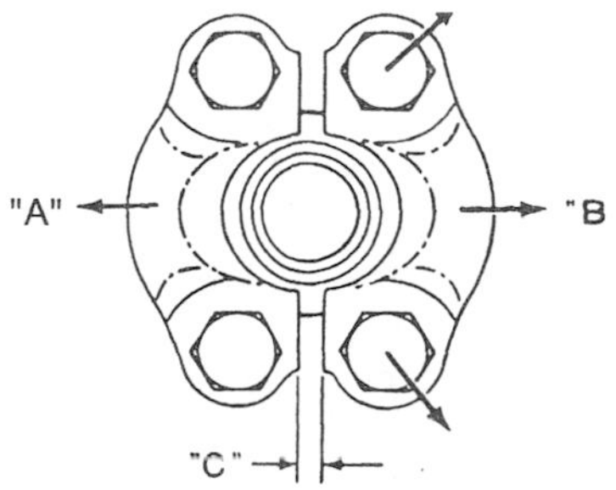

Figure 6.32 Tolerance buildup can minimize holding power of the 4-bolt split flange.

greater with the above conditions (Figure 6.33). Make certain flanges have a small break at edge "D" to insure full contact with the shoulder (Figure 6.34).

In spite of all of the care required for use of these flanges they have provided good service with minimal trouble when properly installed. Mobile and utility maintenance vehicles have many years of satisfactory service in the most hostile environment in which they must function with minimal maintenance.

The flange of Figure 6.35 employs a 15 degree port with matching nose on the flange. The flange is fitted with an elastomeric seal to backup the basic metal-to-metal seal (NWD International, Inc.; patented).

OTHER CAUSES OF LEAKS

Many other conditions can cause external leakage. Vibration (Chapter 5) that can loosen connections is obviously a major culprit. In addition, the troubleshooter should consider the possibility of improper piping assembly, where fittings were not tightened properly in the first place as previously outlined.

Damaged or worn seals (in linear as well as rotary components) are a common cause of leakage. While replacing defective seals, the troubleshooter should also consider steps

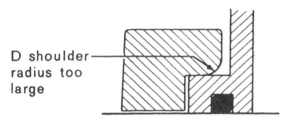

D shoulder radius too large

Figure 6.33 An excessively large radius can minimize holding power.

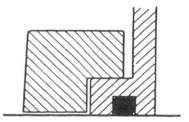

Figure 6.34 Square corners with minimum corner break can enhance holding power.

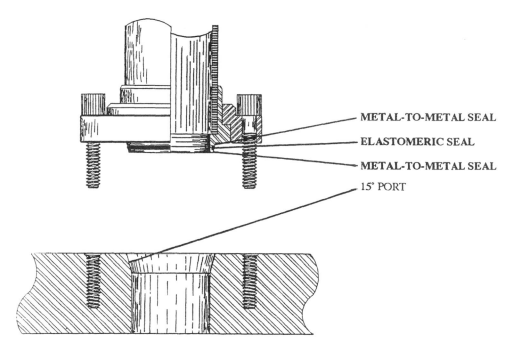

Figure 6.35 Metal-to-metal contact is provided in the 15 degree flange with an elastomeric seal to insure a leak-proof connection (NWD International, Inc.; Patent No. 5115550). Courtesy of NWD International, Inc.

that might be taken to eliminate—or at least minimize—seal leakage in the future. For instance, are seals made of the appropriate elastomer? Might a different seal design reduce leakage? Should cylinder rods be specified with different surface finish? Would boots help to extend seal life and thus reduce leakage?

Such thoughts should be communicated to the company's own top management who in turn should review these factors in the future when the company plans to buy additional machinery powered by hydraulics or pneumatics. Good planning before purchasing might help eliminate the necessity for a lot of troubleshooting.

7

System Tests for Gear and Vane Pumps

This troubleshooting guide has dwelt on a 2-step process: logical mental analysis, followed by hands-on investigation and correction. We have considered the three major symptoms—noise, heat, and vibration—of actual or impending hydraulic equipment malfunctions and looked at ways to track down their causes. Now we turn to analysis of problems exhibited by specific components.

A hydraulic pump is the most important component of a major piece of construction equipment. It may also be the most important component in a machine tool or automated manufacturing system. It operates far more hours than any other component, yet is often the most neglected component.

Hydraulic pumps are precision instruments that are manufactured to watchmaker tolerances. So only with careful attention to the things that will damage these clearances can accelerated wear, pump inefficiency, or unexpected breakdown be prevented. A lack of preventative maintenance or operator abuse invariably will lead to costly pump replacement or reduced equipment availability.

Operators of the equipment as well as maintenance personnel must learn to recognize symptoms of impending trouble. Pump problems often mean that something else is wrong.

When a hydraulic system has problems, it is common practice to first look at the pump. The pump is arguably the heart of almost every hydraulic system and, as such, has traditionally been blamed for almost every operating problem. Obviously this cannot be true, but when preliminary troubleshooting analysis focuses on the pump, a series of steps can pinpoint the specific problem and lead to a cure.

Major causes of pump trouble are hydraulic shock, oil contamination, excessive heat and pressure, oil cavitation, and operator abuse. Since one symptom may cause or aggravate another, the alert hydraulic maintenance person will not be satisfied merely to replace a damaged pump. He/she will look for and correct a problem elsewhere in the system, or will bring to managements' attention evidence of operator abuse or negligence that may have caused the difficulty.

Hydraulic shock is a major contributor to hydraulic pump failure. A major source of shock in a hydraulic system could be a damaged or improperly set relief valve. For example, if a relief valve is slow in opening, excessive pressures are produced, virtually instantaneously. Then it is possible for the internal pressure of the system to exceed the

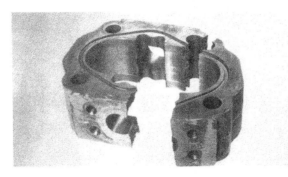

Figure 7.1 Excessive shock fractured this housing either because of a faulty or improperly set relief valve. Courtesy of Commercial Intertech.

design relief pressure limits, sometimes reaching the fracture point of some part of the circuit or the pump itself (Figure 7.1).

Other parts of the system may cause problems with the relief valve such as an undersized elbow or hose or an incorrect fitting downstream from the valve that restricts flow and completely alters the performance characteristics of the relief valve. Fast reversal of a directional control valve and/or stopping a heavy load can put additional strain on the relief valve and the pump, particularly if the relief valve has been adjusted above design pressure values.

Operating characteristics can also be changed by tampering with the relief valve adjustment in an attempt to improve pump performance. An operator may increase tension on the bias spring in the relief valve under the mistaken impression that he or she is improving the performance of the machine by increasing hydraulic system pressure. Obviously what is happening is an increased heat buildup, increased risk of shock, and practically guaranteeing premature pump failure.

Resetting a pump governor is another way an operator may attempt to improve system performance. This is not an approved technique either, since it results only in heat buildup and premature pump failure.

Excessive work by the pump always translates into wear, and this will eventually cause failure. Further, the worn pump is replaced by a new pump with its close factory tolerances and if the relief valve and governor are not checked and reset, the same conditions which caused the first pump to fail will overload the new pump from the start. Also, a new pump with its close clearances may fail much more quickly than the older worn pump. Many complaints are received with the user stating that the new pump failed only after a few hours while the original pump lasted many years.

The inability of a relief valve to perform properly may cause a fitting or hose to fail immediately. Drive shaft or gear housing fracture sometimes can occur immediately after installation of a new pump (Figure 7.2).

Recommended maintenance procedure after installing a replacement pump is to back off the relief valve setting and let it run in for half a minute. It should be operated at low load for two or three minutes and then, following the equipment manufacturer's instructions, brought up to full pressure.

This technique will also help to bring the pump to operating temperature gradually,

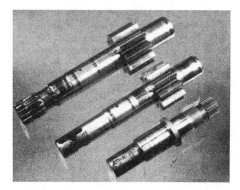

Figure 7.2 Top shaft in this picture was mis-aligned and damaged by coupling. The second shaft was ruined when key was left out when shaft was lined to drive. The third shaft was loaded above safety limit. Courtesy of Commercial Intertech.

Figure 7.3 Contaminated oil forced outward by gear teeth eroded this plate. Courtesy of Commercial Intertech.

especially if the rest of the system is already hot; hot oil should never be introduced into a cold pump, especially during the winter or under arctic conditions.

Excessive pressure in the hydraulic system can be caused by overloading the hydraulic system or by operating it at speeds and horsepowers beyond the manufacturer's recommendations. In case of doubt or when using attachments or counterweights for which the equipment was not designed, consult with the vendor's engineering department to make certain the hydraulic pump and hydraulic system can handle the extra load.

Added pressure greatly increase the chances of leakage at fittings, swivels, shaft seals, gland packing, and gaskets. Whatever the cause, any leakage results in lower machine efficiency and increased heat generation. Heat buildup, in turn, causes hydraulic oil to lose its viscosity resulting in an oil that no longer meets specifications and which oxidizes, encouraging corrosion, sludge formation, and excessive leakage.

The deterioration of a hydraulic oil that is operating too hot greatly reduces the lubricating effect on high tolerance surfaces; this causes high loadings and high frictional contract, and results in accelerated wear on these surfaces as well as on pump parts and clearances (Figure 7.3).

Remember that pumps do not "pump pressure"; they only deliver hydraulic fluid from the reservoir to the system. Pressure at a pump outlet is the result of some downstream resistance to flow (Figure 7.4). This is true of all pumps, but not all pumps generate flow in the same manner. We will begin our investigation of pumps with the two simplest designs: gear and vane pumps.

THE GEAR PUMP

The gear pump is produced in several configurations. The teeth of the spur gear pump of Figure 7.5a unmesh as the shaft rotates. A vacuum is created by the unmeshing action which is satisfied by atmospheric pressure on the liquid in the reservoir urging it into the

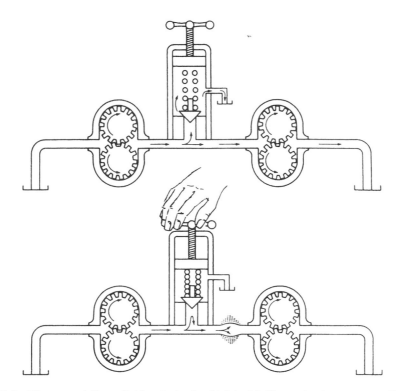

Figure 7.4 The pump delivers fluid to the load, which in this illustration is a gear-type fluid motor. Pressurized fluid will be diverted to the reservoir if the resistance to turning of the fluid motor increases pressure level to the value set at the relief valve. Turning relief valve adjustment without suitable reference to gauge or circuit information can cause rupture of weakest portion of power-transmission components.

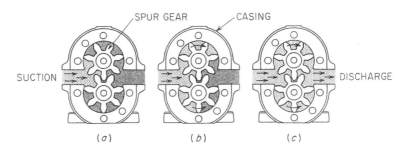

Figure 7.5 Rotary spur gear pump. (a) Atmospheric pressure urges fluid into the pump inlet. (b) Cavities between the teeth carry the fluid to the outlet port. (c) Fluid is discharged at a pressure resulting from the resistance to flow at the outlet port.

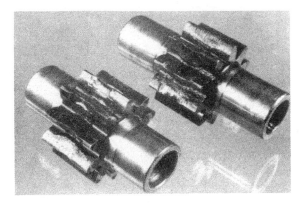

Figure 7.6 A large chunk of foreign material passed between these gear teeth, damaging them severely. Courtesy of Commercial Intertech.

suction pipe and into the pump housing. The liquid travels in the pocket between the teeth around the housing to the discharge (Figure 7.5b). As the teeth mesh the displaced liquid is forced out the discharge port (Figure 7.5c).

The spur gear pump must discharge the fluid as the tooth meshes and fill the cavity as the tooth unmeshes. This must happen in approximately seven degrees of shaft rotation. The ability to flow at this rate limits the maximum effective rotating speed of the spur gear pump. Foreign matter in the fluid medium can severely damage the mating gears (Figure 7.6). The crescent-type gear pump of Figure 7.7 can fill for one-third or more of the shaft rotation and discharge over one-third or more of the shaft rotation. Consequently it can operate at a higher rotating rate than the spur gear pump.

The generated rotor-type gear pump of Figure 7.8 can fill for approximately 180 degrees and discharge for approximately 180 degrees so that it can function at a higher shaft rotation speed than other types of gear pumps. The generated pinion and ring gear format eliminates the need for the crescent element.

THE VANE PUMP

The unbalanced vane-type pump design of Figure 7.9 operates at relatively low pressures. The rotor is eccentric to the reaction ring within the pump housing. The balanced vane-type pump design employs the equivalent of two pumping actions. Each pumping action uses one-half of the displacement assembly. As the vanes enter the upper eccentric lobe fluid is directed to the pockets between the vanes. The fluid is carried to the part of the ring where the vanes are forced to move into the slot and the fluid is directed to the discharge of the pump. An exact replica of this action occurs in the lower segment of the vane, ring, and rotor assembly. Thus each rotation is involved with vanes in the upper ring area and the other half of the rotation is involved with vanes in the lower ring area. The balanced vane pump can operate at much higher pressures with minimum shaft loading because of the balanced radial forces. The sectional view of Figure 7.10 illustrates the relative position of the ports. The end view of Figure 7.11 illustrates a pump with dual vanes in each slot. The vanes are chamfered to provide the balance areas so that centrifugal force and hydraulic

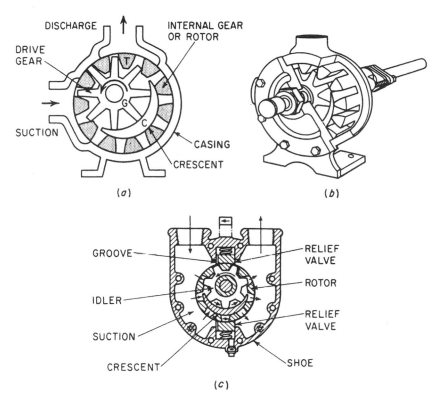

Figure 7.7 Internal gear pump. (a) The pinion gear is eccentric to the ring gear. The crescent fills the space and acts as a seal element in conjunction with the housing. (b) Cutaway view of internal gear pump. (c) A relief valve function can be incorporated in the pump housing.

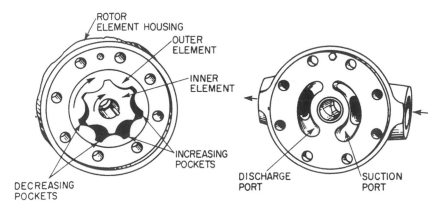

Figure 7.8 Generated rotor pump eliminates need for the crescent divider.

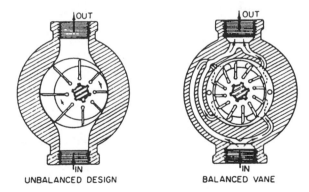

Figure 7.9 Unbalanced vane pump fills rapidly. Balanced vane pump includes two input ports to the vane and rotor sector and two discharge ports so that the radial forces are balanced and minimum shaft deflection occurs.

forces employed to keep the vanes in contact with the ring are minimized for longer service life with higher pressure capabilities.

PUMP MALFUNCTIONS

The most common pump malfunctions are inadequate delivery of flow or an inability to develop required system pressure. As a matter of general interest, the failure mode differs between vane and gear pumps. Most often, gear pumps lose pumping efficiency over a

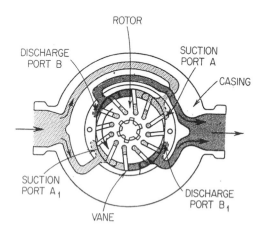

Figure 7.10 Typical porting arrangement for a balanced vane pump.

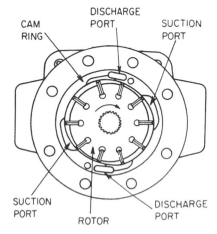

Figure 7.11 Two chamfered vanes may be provided in each slot to minimize wear from forces employed to hold the vanes in contact with the reaction ring.

period of time and fail *gradually*. Probably the equipment operator has noticed this slow deterioration of pump performance. Vane pumps, on the other hand, are more apt to stop pumping *suddenly*, with little or no advance warning.

THE HUMAN ELEMENT

The human senses of hearing, touch, and sight play an important role in analyzing malfunction. The ear can easily detect the rumble of bad bearings or the distinctive ping of cavitation or entrapped air in the fluid. Cavitation occurs when the pumping mechanism does not get fluid fast enough and localized implosions occur damaging pump walls and gear structure. Our sense of touch can quickly identify excessive heat and determine if it is localized at the pump. The eye can note which components are moving and which are not— and also read the pressure gauges installed in the system. A gauge at the pump outlet is essential for proper evaluation and should be added to every hydraulic machine. A pressure gauge isolater valve can be added to reduce wear on the gauge. The symbol is shown with pressure normally blocked and the gauge connected to tank (Figure 7.12). Pushing the operator button against the bias spring directs pressurized fluid to the gauge for the pressure level reading. When the button is released the bias spring returns the spool to a position to block pressure and relieve the gauge to the reservoir line. An alternate is the portable pressure meter that connects to special couplings permanently installed in the piping system. At a minimum a needle valve should be installed permanently with a plug to protect the line from foreign matter. When needed a gauge can be installed without any time loss or need to modify the piping.

FOLLOW THE CHART

The chart provides a systematic sequence of troubleshooting steps beginning with a reading of that gauge (Figure 7.13). No pressure indicates no flow. The left side of the chart addresses this situation. Note that most of these checks can be made without instrumentation. (Also note that the item on the pump assembly would be a consideration only on new installations.)

If the gauge at the pump outlet registers pressure, the steps in the right-hand area of the chart should be followed. These steps require some instrumentation (e.g., vacuum gauges and flow meters). Note here that the relief valve and its setting may be at the root of many problems attributed to the pump.

Figure 7.12   Symbol for gauge cutoff valve. Pressure is normally blocked and gauge is connected to tank. Depressing the button directs pressure to the gauge and blocks the tank connection. Release of button blocks input flow and directs gauge to tank connection.

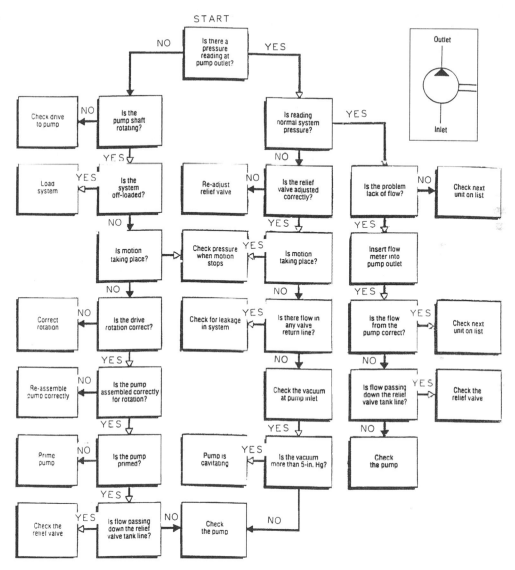

Figure 7.13 Troubleshooting chart for gear and vane pumps. Courtesy of Vickers, Incorporated.

THE FLUID MEDIUM

Understanding the manufacturer's recommendations for oil viscosity and for the correct selection of hydraulic oils is essential for maintaining the service life of pumps and system components. The manufacturer specifies oils and hydraulic system parts which are compatible, and a contractor's maintenance personnel revises them at their peril. For example, certain oil additives attack some kinds of seals; others can pick up water, air, or contaminants that may shorten the service life of components or damage parts machined to precision tolerances.

"Match, don't mix" is an excellent rule to follow in selecting, storing, and using all hydraulic oils, crankcase oils, and lubricants for any hydraulic machinery and especially construction machinery. One way to avoid difficulties is to color-code containers and their closures in an effort to prevent unwanted mixing.

Hydraulic oil contamination is the single, most common cause of hydraulic pump failure. The entry of dirt into the hydraulic system *must* be prevented. All storage tanks, transfer pumps, and fittings should be maintained scrupulously clean. The manufacturer's recommendations for oil and filter changes should be followed meticulously.

Even under normal usage, oil has a definite service life and it should be routinely replaced before it has lost its characteristics of film strength, viscosity, and lubricity. Samples taken in approved containers at intervals can be sent to laboratories for analysis and recommendation as to additive needs or perhaps a complete change. Severe operating conditions can cause accelerated breakdown of a hydraulic oil and these should be taken into account in a maintenance program.

Regular maintenance checks of a hydraulic system will head off surprises. Oil levels should be checked. Leaks should be looked for at fittings, swivels, and connections, as well as at all seals, packings, and gaskets.

The possibility of a missing screen or collapsed filter element should be considered, as well as a lost or loose filter cap. A breather becoming loose or lost can admit dirt to a system, particularly as the oil level drops when cylinders are filled during operation. In this event, both the oil and the filters should be changed.

CLEANING THE TANK

Flushing the tank and system with straight kerosene is usually bad news. Kerosene is not compatible with most hydraulic fluids and even small amounts left in the system can immediately contaminate the clean oil. Avoid flushing with solvents or chemical cleaners. Unless specifically recommended by the machine builder, most of these products will damage seals and hoses.

It is bad practice to add kerosene or diesel fuel to hydraulic oils in order to thin them. They are not sufficiently refined for hydraulic systems. Kerosene or diesel oil will change the aniline point of the oil and may soften or swell the seals. Use only the oil rated for your type of operation.

INSPECT FILTERS REGULARLY

Many hydraulic systems include a bypass which permits oil flow when a filter becomes clogged. The hydraulic pump continues to operate and the oil then carries contaminants to all parts of the system. Unless the condition is remedied promptly the pump and other components are subjected to accelerated wear.

Contaminants dispersed in the hydraulic oil score and damage thrust plate surfaces (Figure 7.14). Thrust plates are held against the pump gears by discharge pressure providing a bearing surface and a seal against the gears, but as erosion continues, pump efficiency drops.

Wear at bearing contact areas is usually caused by contamination. While roller bearings are more resistant than bushings to contaminant wear, anything that wears can put abrasive particles into suspension which can destroy other parts in the system.

Figure 7.14 Coarse grit ruined both the pressurized wear plate and the pump into which it was fitted. Courtesy of Commercial Intertech.

LOW OIL LEVEL INVITES CONDENSATION

Low oil level is a serious cause of contamination in the hydraulic system. When the oil level is low, more air gets into the tank and this often leads to destructive cavitation in the pump and to condensation on the tank walls which generates sludge. Sludge decreases the lubricity of the oil, producing scoring and friction on surfaces with close tolerances.

Cavitation is the result of implosion—breakdown—of air or oil-vapor bubbles in the hydraulic system. It can be prevented by making certain that suction screens (if used) are clean and the oil reservoir is always full; this is another good reason for preventing leakage anywhere in the system. A clean reservoir with good return line filtration may eliminate the need for suction screens. Suction screens can be difficult to clean and as a result they may not be cleaned and pumps may be lost as the screens plug.

USE THE RIGHT OIL

It is obvious that dirty oil causing system failures is expensive. The wrong oil can be expensive, too. Four things affect your choice of oil:

1. the type of equipment
2. the manner in which the equipment is operated
3. the climatic conditions
4. the type of seals and hydraulic components

The best advice on the right oil to use can be given by:

1. the original equipment dealer who represents the manufacturer
2. the oil supplier

Some types of hydraulic components require less exotic additives than others. In the long run, your experience may prove the most practical guide.

Correct Viscosity

At operating temperature, the oil should have viscosity of between 100 SSU and 50 SSU (Saybolt Seconds Universal) for gear and vane pumps. You should not run your system with oil thinner than 50 SSU.

At high operating pressures, oils thinner than 50 SSU will slip through clearances built into hydraulic components, resulting in efficiency loss. Thin oils lose their film strength and cause localized overheating. As a consequence of heat buildup, steel parts may show bluing.

Oils that are too heavy will not adequately supply the pump. This can cause cavitation damage, overheating, and sluggish operation.

Viscosity Index

Oils with a high viscosity index (90 or above) are recommended for hydraulic systems. The viscosity index of an oil is an indication of viscosity change due to temperature variation. Low viscosity index oils thin out too quickly with rising temperatures.

ADDITIVES

Oils rated for hydraulic service generally contain additives which improve the viscosity index, inhibit rust and oxidation, and often include a foam depressant.

ANILINE POINT

The aniline point should never be below 175, and 200 or higher is preferred. If the aniline point is below 175, the seals may fail and leak prematurely, and cylinder packings may swell up and bind on the tubes.

Oils for Winter Use

To minimize warm-up time during cold morning start-ups, you need an oil that will flow at the lowest expected temperature. Oil viscosity at start-up should never exceed 7500 SSU. Pour-point should not be less than 20 degrees below anticipated outdoor temperatures. Along with winter-service hydraulic oils, there are other types of oils that may be considered (e.g., SAW 10W motor oils, multiple viscosity oils such as 10W-20, and ATF or Automatic Transmission Fluid oils).

Cold-weather operation can not be avoided, but common sense and a few precautions can minimize downtime. These three basic steps will help assure satisfactory operation under the worst conditions.

1. Use equipment rated for the conditions anticipated.
2. Closely follow the manufacturer's recommended start-up, operating, and preventive maintenance procedures.
3. Use the correct cold-weather oil.

Proper Warm-Up is Vital

Seals used in standard components of hydraulic systems are generally made of nitrile (Buna N) compounds. These seals work quite well, even at arctic temperatures. They may become brittle at low temperatures, but proper warm-up procedures can protect them from mechanical shock. Special low-temperature seal compositions are available, but they lose high-temperature compatibility and tend to be less oil resistant.

Do Not Load a Cold System

Keep speed down and do not load the system until it is warm. Pump seizure, with possible internal scoring or shaft breakage, can occur quickly with cold oil. Gear pumps will withstand a certain amount of cavitation on start-up. There will be no cavitational damage as long as the pump is not pressurized and there is enough oil for lubrication.

Operate all the various functions without loads to circulate the oil throughout the system. To shorten warm-up time, the oil may be eased over the relief valve occasionally. *Caution:* Do not force oil over the relief valve if you hear or suspect that the pump is cavitating. Reservoir and line heaters are valuable accessories in extremely cold climates.

Never Feed Hot Oil into a Cold Pump

During an interim shutdown, an exposed pump may cool down more than the oil in the reservoir. Start-up should be done only after warming the pump externally. Jogging may work, but allow extra time or the pump will seize.

All-Season Oils

Some oil companies now offer all-season hydraulic oils. They contain viscosity index improvers and pour-point depressants. Some automatic transmission fluid oils also can be used year-round.

Multipurpose Oils

Several oils on the market are rated as multipurpose. Although slightly more costly, they can simplify inventories and reduce service downtime. *The wrong oil, like dirty oil, can be very expensive.*

WATCH OUT FOR WATER

Avoid water condensation and accumulation. Cold-weather operation invites condensation in the system. Water must be regularly vented from the tank and from the lowest parts of the system.

Water damages the oil, destroys its lubricity, forms sludges, causes rusting, and freezes solid. Protect your idle equipment from falling snow and fluctuating temperatures with a shelter or tarpaulin. Scheduling equipment for longer periods of operation will discourage condensation. When moving equipment indoors, remove snow and, if possible, operate it long enough to evaporate condensation.

WATCH THE MAINTENANCE SCHEDULE

Cold-weather operation necessitates more frequent oil changes and flushing of the system. Have oil tested at a qualified laboratory to determine if additives and/or a complete oil change are needed. Make frequent checks of the system following the manufacturer's instructions.

KNOW THE CAUSE OF PUMP FAILURES

Knowing the causes of pump failure can lead to corrective action before actual breakdown. Once the causes are recognized and acted upon by operators and maintenance personnel, premature pump failure can be prevented.

8
System Tests for Piston Pumps

Although the troubleshooting chart in this chapter is focused on pressure-compensated, variable-displacement piston pumps, it can be used to check any type of variable-displacement pump such as the variable-displacement vane pump of Figure 8.1.

COMPENSATOR SETTING

Adding pressure-compensating capability to a pump creates a new and important factor to be considered when troubleshooting a circuit. The pressure compensator monitors downstream pressure and automatically changes the pump's internal configuration to alter flow so as to maintain a preset pressure at the outlet of the pump.

A common problem in circuits with pressure-compensated, variable-displacement pumps is incorrect setting of the compensator cut-off pressure in relation to the spring setting of the system relief valve. The relief valve must always be set higher—typically 100–150 psi—than the compensator. The reason is that if the compensator is set higher, part or all of pump output will flow over the relief valve when system pressure reaches the relief valve setting. The compensator will, in effect, be bypassed and the operating concept for which the system designer specified pressure compensation will be negated. In addition, fluid temperature will rise due to the energy loss resulting from the pressure drop over the relief valve.

CASE DRAIN

Another important consideration in these circuits is case drain pressure, which is usually in the range of 5 psi. In many pump designs, the pressure compensator drain flows into the case, allowing pump case pressure to affect the compensator (Figure 8.2). Should excessive case pressure develop from a partially obstructed case drain line, for example, this additional back pressure could cause serious problems. An erratic, fluttering compensator could quickly destroy the pump mechanically.

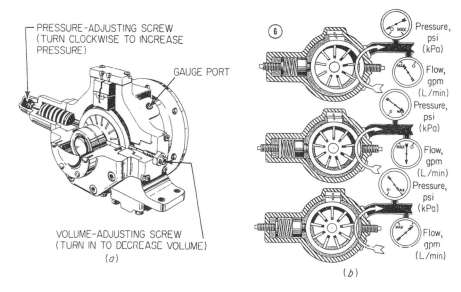

Figure 8.1 Pressure-compensated variable-displacement vane pump. (a) Bias spring urges reaction ring to maximum flow position. (b) Maximum flow occurs with minimum pressure. As pressure rises flow decreases. At established pressure level the flow decreases to that needed to make up for lubrication and clearance flows. Courtesy of Racine Hydraulics and Machinery Company.

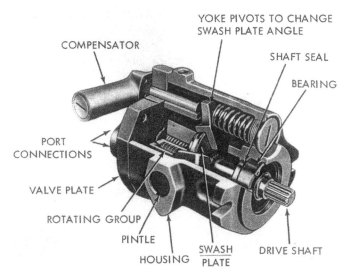

Figure 8.2 Variable-displacement axial piston pump. Note compensator discharge into the pump housing. Courtesy of Vickers, Incorporated.

INCORRECT PRESSURE

Again, a pressure gauge installed downstream from the pump is a most valuable troubleshooting tool. If the gauge shows no pressure, follow the checkpoints in the extreme left-hand column of the chart (Figure 8.3).

First, note if the pump shaft is rotating and in the correct direction. Next establish whether lack of system pressure might be caused by an "open system" condition: no resistance to fluid flow because of an open directional control valve or an incorrectly set relief valve. If the pump is not primed, an air-bleed valve installed downstream will help the pump self-prime (Figure 8.4). Air-bleed valves are available in cartridge format to minimize piping. Air will pass readily through the orifice in the valve. As the more viscous oil reaches the valve the resistance will cause the spool to close against the bias spring, thereby blocking the return to tank until pressure is released when the system is shut down.

If the system is developing some pressure—but less than desired or required—follow

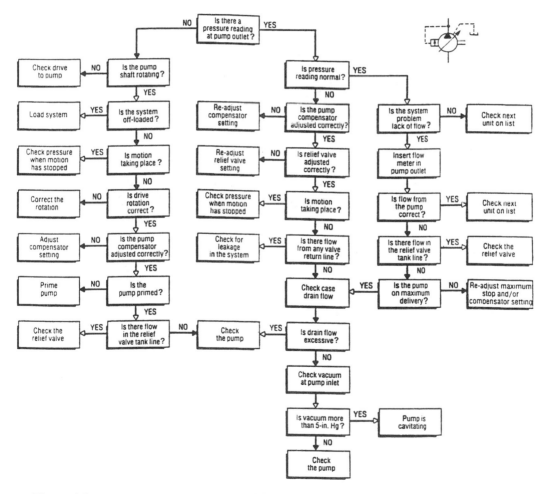

Figure 8.3 Troubleshooting chart for variable-displacement pumps. Courtesy of Vickers, Incorporated.

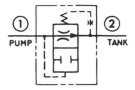

Figure 8.4   An air bleed valve will pass air but will not pass solid oil. Courtesy of Sun Hydraulics Corporation.

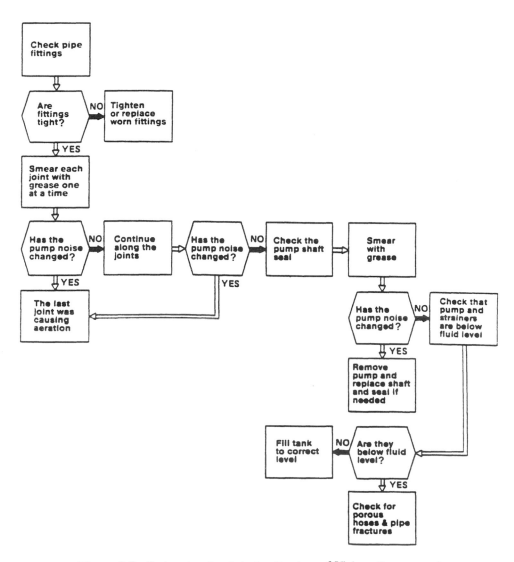

Figure 8.5   System test for air leaks. Courtesy of Vickers, Incorporated.

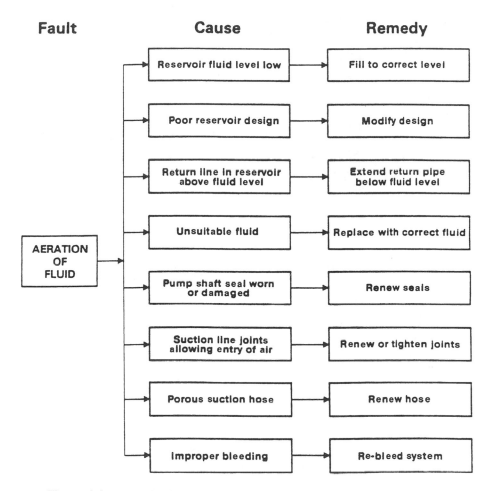

Fault   **Cause**   **Remedy**

Figure 8.6 Remedies for aeration of fluid. Courtesy of Vickers, Incorporated.

the check points on the extreme heating, erratic performance, and possible damage to the pumping mechanism. Follow the chart of Figure 8.5 to find air leaks in a pump suction line.

A flooded suction usually results in an external leak if the pump is not operating making it easier to find an air leak in the suction line. Minor leaks may not be obvious until the pump is operating. The tests of Figure 8.5 should be performed with the pump operating.

AERATION OF FLUID

Air within a hydraulic circuit can create a spongy response, a noisy pump and possibly cavitation. The chart of Figure 8.6 outlines potential causes of aeration and suggests remedies.

PUMP CAVITATION

The chart of Figure 8.7 outlines potential causes of pump cavitation and suggests remedies.

Fault   **Cause**   **Remedy**

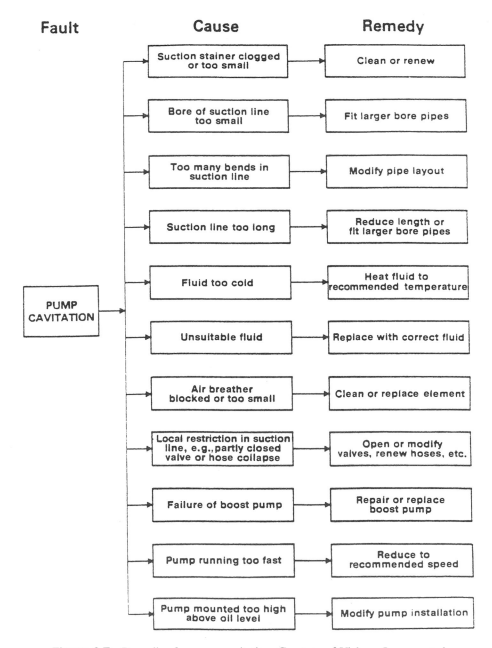

Figure 8.7 Remedies for pump cavitation. Courtesy of Vickers, Incorporated.

9
Relief Valves

The often unsung hero of many hydraulic systems is the relief valve. The contribution of the hydraulic relief valve is extremely important because it limits the amount of power a system can generate. The relief valve thus helps to ensure that intended system pressure is not exceeded. This pressure limitation helps avoid damage to other hydraulic components and to the machinery, or injury to the workers.

The relief valve function helps establish the fact that hydraulics is a convenient and efficient method for transmitting power. Mechanical and electrical power systems cannot match the simplicity and low cost of the relief valve.

THE RELIEF VALVE OPTIONS

A relief valve is basically a normally-closed two-way valve that is biased to the closed position by a load, a spring, and/or a spring and pressurized fluid. As the line pressure at the inlet to the relief valve approaches the force value created by the bias mechanism an orifice is created between the pressure input and the line to the reservoir. The size of the orifice is that needed to pass the pressurized oil at the rate at which it is being supplied by the pump and/or other sources at the desired maximum pressure level.

A safety relief valve may be nonadjustable (Figure 9.1) with the seated area of the valve biased by the mechanical spring. It may be a two-diameter spool type assembly with the difference in diameter biased by the mechanical spring (Figure 9.2). A pilot plunger may be used to reduce the effective area of the forces moving the piston against the bias spring (Figure 9.3). The closure between pressure and tank can be a simple ball held against an appropriate seat (Figure 9.4).

As pressures and flows increase the versatility of the hydraulic pilot structures increase in importance. The relief valve of Figure 9.5 illustrates the principal of operation of the pilot structure. The main spool is biased to the closed position by the larger spring at adjustment A. Typically it would not be adjustable and might create a pressure between 50 and 100 psi. A source of pilot control oil is supplied from the pressure input through a nonadjustable orifice to the control chamber to assist the bias spring urging the main spool to the closed position. The spring at adjustment B urges the pilot cone against the seat

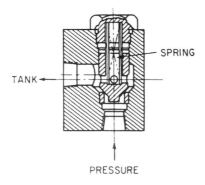

Figure 9.1   Nonadjustable safety relief valve.

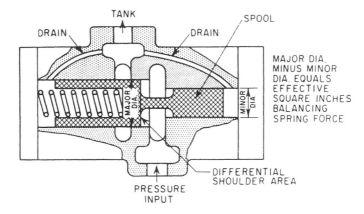

Figure 9.2   Nonadjustable safety relief valve using the difference between two diameters to urge the piston against the bias spring to pass fluid to the tank port.

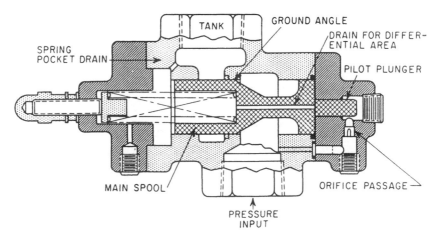

Figure 9.3   Pilot plunger urges the piston against the bias spring to create a passage from input pressure to tank.

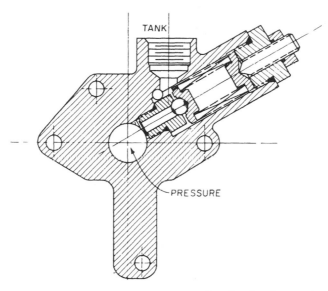

TANK

PRESSURE

Figure 9.4 Adjustable relief valve to sandwich into directional valve stack using a ball spring biased against a seat.

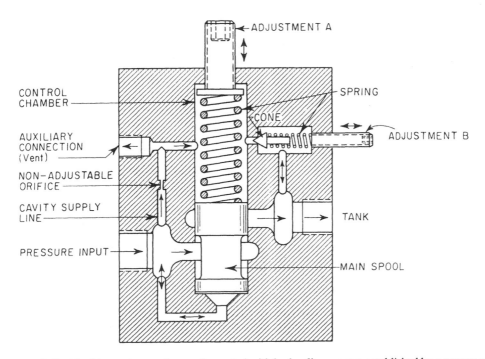

ADJUSTMENT A

CONTROL
CHAMBER

SPRING

CONE

AUXILIARY
CONNECTION
(Vent)

ADJUSTMENT B

NON-ADJUSTABLE
ORIFICE

CAVITY SUPPLY
LINE

TANK

PRESSURE INPUT

MAIN SPOOL

Figure 9.5 The bias spring can be supplemented with hydraulic pressure established by a source of pilot pressure and an auxiliary pilot relief valve.

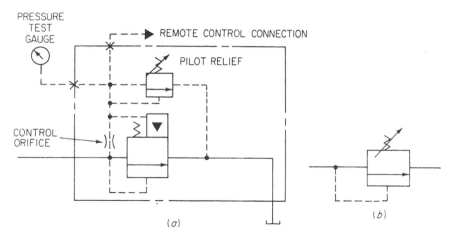

Figure 9.6 (a) Complete symbol for the compound relief valve illustrating the pilot function. (b) Simplified symbol for a compound relief valve.

adjacent to the control chamber. The main spool will remain closed until the pressure input level reaches the value set at adjustment B. When the pilot oil escapes past the cone faster than it can be supplied through the nonadjustable orifice the main spool becomes unbalanced creating a passage from the pressure input to the tank. Many different main spool designs will be encountered but the principle of operation will be similar.

The compound relief valve of Figure 9.5 is shown in symbolic form in Figure 9.6a. Note the equilateral triangle adjacent to the main body indicating the hydraulic force potential. The remote control connection offers the option of control by a small pilot valve remote from the piloted valve. A simplified symbol is shown in Figure 9.6b. Figure 9.7 illustrates three optional remote pilot valves that can be employed to provide the adjustments used to control the pressure level of the piloted valve. Pilot valves connected in series or parallel to the control chamber of the piloted valve will provide an adjustment at a remote

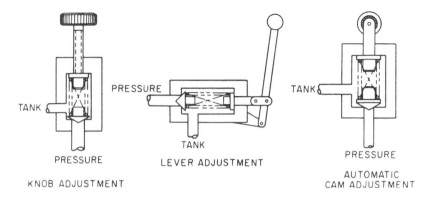

Figure 9.7 Typical adjustment options for remote control of the basic relief valve.

point. The pilot in parallel with the lowest setting will control the maximum pressure value at the piloted valve. Pilot valves in series will respond to the valves in a cascade mode. The last pilot in the series will be additive to each of the upstream valves. Thus a remote pilot connection to a knob-adjusted valve could establish a pressure level that might be increased to higher values by a cam- or lever-operated remote pilot valve in series. The pilot in the main relief valve can be set for a maximum desired value that cannot be overridden by the other pilots in a parallel circuit.

A working knowledge of the piloting function of pressure control valves is essential for troubleshooting. From these examples of typical pilot valve circuits it is obvious that any one of the remote pilot adjustments can affect the pressure level established by the piloted valve. A malfunction of the pilot assembly will defeat any efforts to control pressure level by the piloted mechanism. Foreign matter jamming the spools or poppet assemblies is one of the major causes of pressure loss. The valve is designed to be fail safe. Foreign matter will cause the valve to stick in the open position. Cleaning and resetting will normally solve the sticking condition. Obviously, the source of contamination must be eliminated to prevent further malfunctions.

CARTRIDGE-TYPE RELIEF VALVES

A typical cartridge-type relief valve with return of the pilot flow internally to the tank port is illustrated in Figure 9.8. An additional port can be provided to permit parallel remote pilot functions (Figure 9.9).

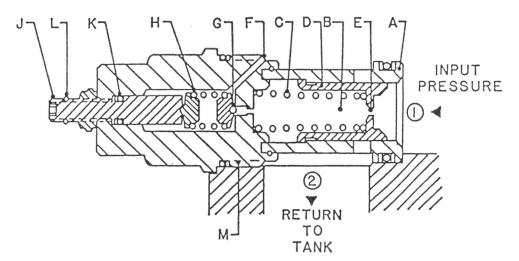

Figure 9.8 Cutaway view of screw-in piloted relief valve in cartridge format. (A) Floating body, (B) control chamber, (C) minimum pressure bias spring, (D) main flow closure, (E) pilot supply orifice, (F) pilot return to tank passage, (G) pilot closure, (H) pilot spring, (J) adjusting screw, (K) seal for adjusting screw, (L) overpressure limiter, (M) thread and static seal assembly. Courtesy of Sun Hydraulics Corporation.

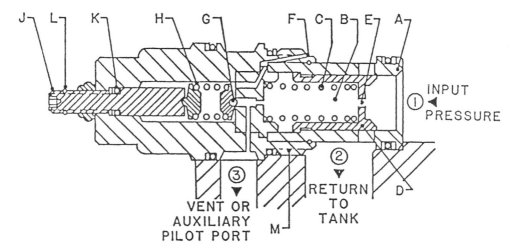

Figure 9.9 Cutaway view of ventable screw-in piloted relief valve in cartridge format with third port for parallel remote pilot control. (A) Floating body, (B) control chamber, (C) minimum pressure bias spring, (D) main flow closure, (E) pilot supply orifice, (F) pilot return to tank passage, (G) pilot closure, (H) pilot spring, (J) adjusting screw, (K) seal for adjusting screw, (L) overpressure limiter, (M) thread and static seal assembly. Courtesy of Sun Hydraulics Corporation.

SYSTEM TEST

Referring to the system test for relief valves (Figure 9.10), the troubleshooter should install a pressure gauge at the up-stream test point—if one is not already in place—to determine the maximum pressure at which the system will operate as controlled by the relief valve. Many types of relief valves provide a port in the valve housing to connect a gauge.

The test procedure chart provides a simple and quick means for determining if the problem is with the relief valve or elsewhere in the system. If the troubleshooter determines that the problem is in fact with the relief valve, Figure 9.11 suggests possible causes of typical troubles.

FAST-RESPONSE RELIEF VALVES

Special purpose relief valves are designed for extremely fast response to avoid pressure spikes when shock loads are anticipated. These special fast-response valves are not intended to create a continuing pressure level. Certain pumps and actuators cannot respond fast enough to changes in the machine function and the fast-acting relief valves prevent undesired pressure levels.

PROGRAMMED ELECTRICAL CONTROL

A programmed pressure level may be desirable. The pilot valve of Figure 9.12 is a small capacity proportional pressure relief valve that is a directly operated, poppet-type valve with electronic remote control. The pressure setting of the valve is directly proportional to

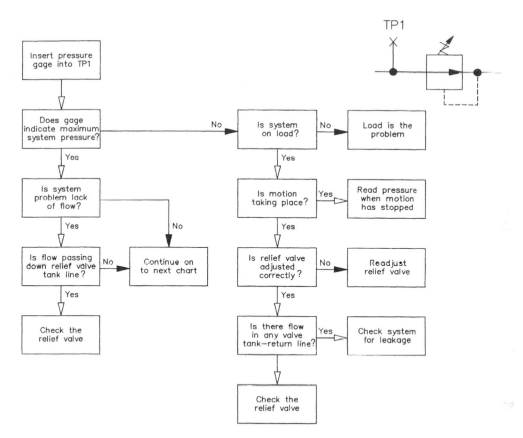

Figure 9.10 Suggested troubleshooting procedures for hydraulic relief valves. Courtesy of Vickers, Incorporated.

Trouble	Probable cause
Erratic pressure	Foreign matter in system Worn poppet/seat in cover Piston sticking in body or cover
Low or no pressure	Valve improperly adjusted Vent connection open Balance hole in piston plugged Poppet in cover not seating
Excessive noise or chatter	High oil velocity through valve Distorted control spring Worn poppet seat in cover Excessive tank line pressure Vent line too long Setting too close to operating pressure

Figure 9.11 Chart of probable causes of relief valve malfunction.

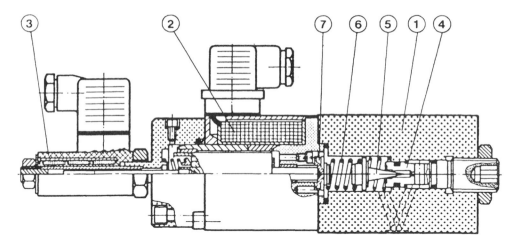

Figure 9.12 Proportional pilot for pressure control valves. (1) Housing, (2) proportional solenoid, (3) linear variable differential transformer [LVDT], (4) adjustable seat assembly, (5) positioning spring, (6) bias spring, (7) pressure pad. Courtesy of Rexroth Corporation.

the electrical input signal. The linear variable differential transformer provides a positional feedback.

The input signal for the pressure setting is given by an input potentiometer (typically 0–9V). This input signal operates via an amplifier and the proportional solenoid (2) to the compression spring (6). The tension in the spring (6)—and the actual position of the pressure pad (7)—is determined by the linear variable differential transformer (LVDT, 3). Any difference between this and the input value is then corrected by the control system. Using this principle, the effects of solenoid friction are overcome. This provides low

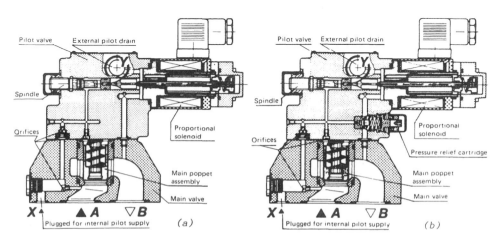

Figure 9.13 Proportional piloted relief valve. (a) Direct control by proportional solenoid without maximum pressure setting, (b) pressure relief cartridge included to establish maximum pressure. Courtesy of Rexroth Corporation.

hysteresis and excellent repeatability. With a zero input signal, loss of current to the proportional solenoid, or break in the feedback cable, the valve will automatically vent to the lowest pressure.

A proportional solenoid can be integrated into a piloted relief valve assembly (Figure 9.13). The valve of Figure 9.13a has a minimum pressure established by the bias spring in the main poppet assembly. A maximum pressure is established by incorporating a small pressure relief cartridge (Figure 9.13b).

Troubleshooting the electronic system follows conventional procedures. Foreign matter in the fluid medium is the most likely cause for malfunction of the pilot mechanism.

THE RELIEF VALVE CANNOT COMPENSATE FOR OPERATOR ABUSE

The service life of any piece of equipment, hydraulic or mechanical, depends on the care you give it. Improper use, which at the time may seem like a work-saving shortcut, puts short-term benefits in question and risks a costly future and perhaps a catastrophe.

Operators can make or break any equipment maintenance program. They can give valuable help by reporting problems early and running equipment within its design limitations, so the operator may be the key element in your service program. The operator is a good start in an educational program to minimize downtime and speed up troubleshooting. You can start by acquainting him or her in operating techniques that protect machines from excessive abuse and a shortened work life.

Improper and abusive operation of equipment should be avoided. Pulses and shocks transmitted through the hydraulic system may be sufficiently cushioned by relief valves, but mechanical shocks can be deadly on equipment that is not designed to absorb them. For example, some backhoes are rated to break pavement or to split rocks, but others aren't (Figure 9.14). Check first to see how your machine is rated before putting it to work. Be sure to use the attachments and accessories approved by your equipment supplier and use them as recommended. Additional relief valve protection may be required with some accessories.

OVERWORK IS HARMFUL

Hydraulic equipment overwork usually means operating at excessive pressures or speeds beyond rated rpm. There is often a temptation to exceed recommended rpm in trying to

Figure 9.14 A relief valve cannot compensate for operator abuse or machine overloading. Courtesy of Commercial Intertech.

generate more power. This is dangerous to both the equipment and the operator. Excessive flow rates and pressures can cause heat build-up and rapid deterioration of hydraulic oil. The result is higher-than-normal maintenance costs and an untimely end for the hydraulic equipment.

Some bad practices that overwork machinery may go unnoticed for a short period of time, unless you check carefully for them. An improperly set relief valve can result in excessive pressures within the system. The same can occur, too, when a governor is not correctly set. Only a good, well-policed program of periodic inspection can protect your equipment against the heavy toll of unskilled operator abuse.

DON'T OVERLOAD THE MACHINE

It's wise to warn operators against overloading. They can't lift a 10-ton load with a 5-ton machine without risk. Adding counterweights and overloading is hazardous. Safety factors are obviously reduced to the point where hydraulic or mechanical failure is probable. There are times, of course, when a machine must be used beyond its normal rating for short periods. In such cases, it is wise to check with the equipment maker to find out how much additional punishment your equipment can take.

Condition your operators to report the earliest signs of wear or excessive heat to avoid inefficient operations. Schedule equipment for service or overhaul regularly or when the signs of wear are first reported. Running machines in less than acceptable condition only speeds wear and adds damage. Replacing the worn part is better than paying for a major overhaul.

10
Pressure Control Valves

Relief valves, sequence valves, counterbalance valves, and unloading valves have one thing in common: They are normally closed and open a flow through the valve as they function in their normal pattern. The major difference is in the drain and pilot function. All will suffer from contamination in the oil, improper pilot source, inadequate pilot source, and improper drain facilities.

SEQUENCE VALVES

In hydraulic systems, sequence valves control fluid flow from the primary to a secondary circuit. A sequence valve performs two functions:

1. It programs the order or sequence in which certain events take place.
2. It maintains a minimum predetermined pressure in the system's primary line while preset events occur in the secondary circuit.

Note that the typical sequence valve's maximum primary pressure setting must not be less than 200 psi *below* the setting of the system relief valve.

To troubleshoot a sequence valve, you need to install pressure gauges in the valve's primary and secondary ports at TP1 and TP2 (Figure 10.1). Some valve manufacturers provide gauge ports in their sequence valve bodies. The pressure gauge readings provide the troubleshooter with the clues needed to understand what is happening within the sequence valve. The symbol indicates that pilot pressure is obtained ahead of the valve and the spring pocket and/or pilot mechanism must be drained to the reservoir.

MULTIFUNCTION VALVES

Economies of design, manufacture, and maintenance have been incorporated in products intended for multipurpose functions within hydraulic circuits. These functions include sequence, unloading and counterbalance. In some circuits a relief function may occur. Thus, certain of these valves may also be used for a relief function. Each function must be

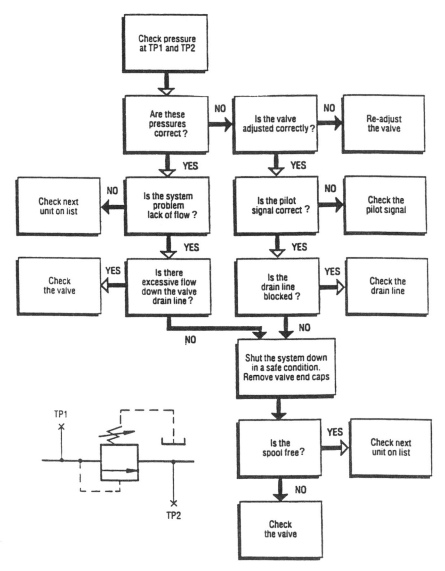

Figure 10.1   Symbol for sequence valve and troubleshooting pattern. Courtesy of Vickers, Incorporated.

identified from the circuit drawing in order to effectively troubleshoot the hydraulic system. The external appearance of the valves destined for the various functions are similar. Differences can be noted by an external pilot connection and an external drain connection. The circuit drawing and bill of material will positively identify the function of the valve and its circuit function.

VARIABLE CAP ASSEMBLY

The sequence valve of Figure 10.2 is assembled with the upper and lower cap positioned so that drain is to an external tank connection. The pilot signal is internal from the input port.

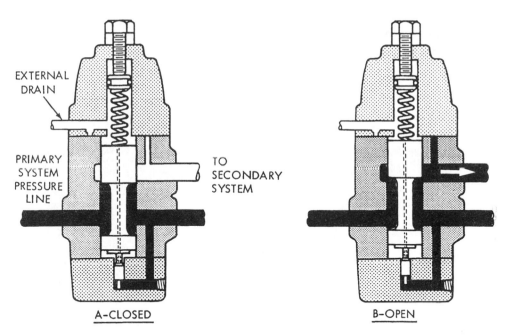

Figure 10.2 Sequence valve assembly. (A) Closed, (B) open. Internal pilot, external drain. Courtesy of Vickers, Incorporated.

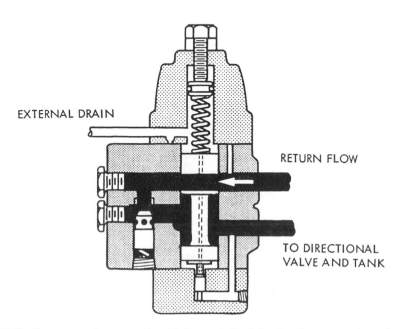

Figure 10.3 Sequence valve assembly with integral check for free flow return through the valve. Courtesy of Vickers, Incorporated.

View A shows the valve in the closed mode and view B shows the flow through the valve as the set pressure is reached. Figure 10.3 illustrates a model with an integral check valve for free flow from outlet to inlet; this is needed if the sequence valve is installed in an actuator line that requires a free return flow. This same valve can be assembled with the cap drain directed internally to the tank port and an external pilot connection (Figure 10.4). The internal drain and external pilot converts the function to that of an unloading valve. A return check is not needed when the valve is installed for an unloading function. A counterbalance function is illustrated in Figure 10.5. The pilot source is internal and the drain is internal. A load is not permitted to move until an appropriate pressure is developed at the pilot connection of the valve.

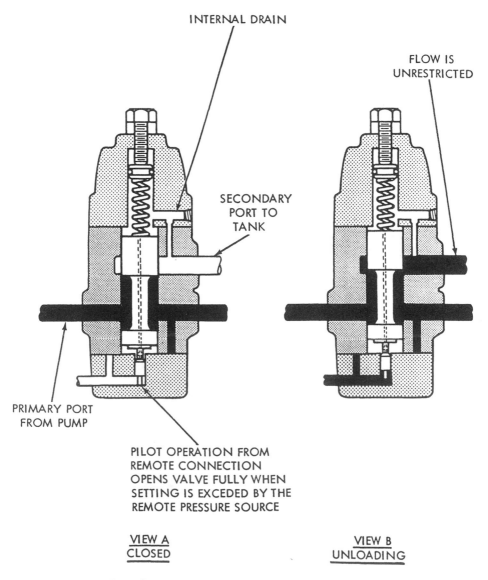

INTERNAL DRAIN

FLOW IS
UNRESTRICTED

SECONDARY
PORT TO
TANK

PRIMARY PORT
FROM PUMP

PILOT OPERATION FROM
REMOTE CONNECTION
OPENS VALVE FULLY WHEN
SETTING IS EXCEDED BY THE
REMOTE PRESSURE SOURCE

VIEW A
CLOSED

VIEW B
UNLOADING

Figure 10.4   Unloading valve assembly without integral check. Courtesy of Vickers, Incorporated.

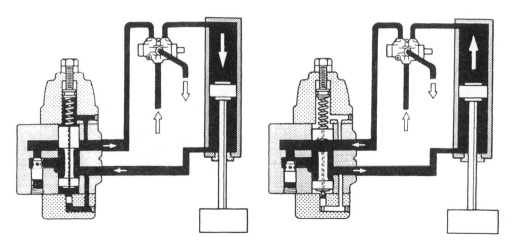

Figure 10.5 Counterbalance function. Internal pilot, internal drain. Load is held until positive pressure is applied. Courtesy of Vickers, Incorporated.

BLOCKING OR UNBLOCKING PILOT PASSAGES

The multipurpose valve of Figure 10.6 can be assembled for sequence, unloading, or counterbalance service by inserting or omitting a plug in the pilot source or pilot drain ports, and with appropriate pilot piping for external drain or external pilot source.

Unauthorized adjustment of multipurpose valves can create a hazard to machinery and personnel. The multipurpose valve of Figure 10.7 is fitted with a locking adjustment. In the locked mode a pin is retracted and the knob spins free. With the key engaged the pin is inserted and the knob can function in the normal manner.

Higher pressures and flows may be controlled effectively with a piloted structure. A counterbalance assembly is shown in Figure 10.8. The symbols of Figure 10.9 illustrate the connections employed to modify the valve structure for counterbalance, sequence, or unloading function.

THE CARTRIDGE FORMAT

The sequence function can be provided in a compact assembly that can be integrated into an actuator, manifold structure, or appropriate housing. Note that the valve in Figure 10.10 is essentially a high performance piloted relief valve with an external drain for the pilot assembly. Input pressure cannot pass to the output port until the pressure value at the input port reaches the setting of spring "H."

SEQUENCE VALVE MALFUNCTIONS

Contaminants in the hydraulic fluid can affect a sequence valve in two ways: by plugging internal pilot passages in the valves and by accelerating wear of all moving parts.

A sequence valve can malfunction in a variety of ways: *Premature valve shifting* (low pressure readings on the gauge at TP1 of Figure 10.1). If oil flows from the primary to the

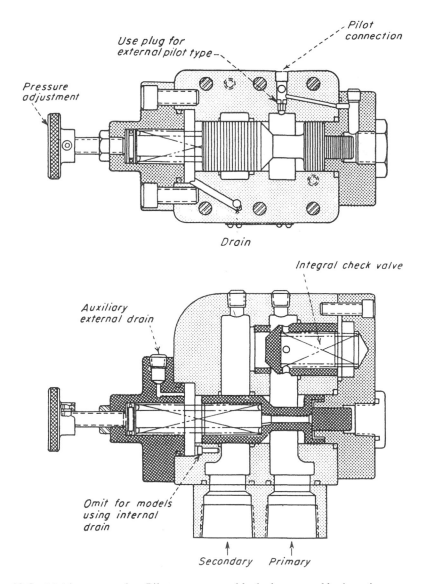

Figure 10.6 Multipurpose valve. Pilot passages are blocked or opened by inserting or removing a threaded plug. Courtesy of Vickers, Incorporated.

secondary port at pressures below the original valve setting, check the sequence valve to determine whether:

- The drain hole through the main spool is plugged (Figures 10.2–6). This would allow pressure to build in the low end of the spool cavity and cause the spool to shift at the wrong time.
- Someone tampered with the original valve setting or it was inadvertently changed.
- The main spool is stuck open because the spool and/or bore are scored or bound with contaminants.
- Pilot orifice is plugged (Figures 10.8–10).

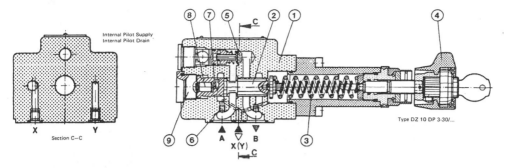

Figure 10.7 Multipurpose valve with locking adjustment. Valve consists of a housing (1), control spool (2), springs (3), pressure adjustment (4), and an optional built in check (5). The pressure at which the valve passes oil is set at the pressure adjustment (4). The springs (3) hold the control spool (2) in the closed position. Pressure in line A acts on the end of the spool (8), which rests against the plug (9) via control passages (6) and (7). If port X is plugged, the signal is given internally from line A via drilled hole (6) and (7) to move the spool. When pressure in line A reaches the set value, the spool moves against the spring to connect port A to B. Oil now passes to the system connected to port B, while maintaining set pressure at port A. The pilot oil may also be supplied externally via port Y or internal with Y line plugged. Check valve (5) is required in order to allow free flow from port B to port A. Courtesy of Rexroth Corporation.

Delayed valve action (high pressure readings on gauge at TP1 of Figure 10.1). If pressure in the primary port rises above the original valve setting, check the sequence valve to determine whether:

- The pilot piston is binding. If so, is it due to scoring or contaminants that keep the main spool from shifting to direct flow from the primary to the secondary port?
- The pilot fluid passage in the valve body is plugged with contaminants thus lowering the pilot pressure source required to shift the spool.

Pressure fluctuations (gauges at TP1 and TP2 of Figure 10.1 show erratic pressure changes) could be caused by:

- changing contamination levels in the valve as contaminants periodically wash through
- varying pressures in the valve drain line to tank.

If this drain line is sized improperly, back pressures may fluctuate as other control functions cut in and out of the system.

No fluid from the secondary port (no pressure at TP2 of Figure 10.1). If the valve drain line is not connected to reservoir or is plugged, back pressure will rise in the top cavity over the main spool (because of internal leakage) to a level where the main spool cannot shift; this is usually encountered in new or rebuilt installations.

THE UNLOADING FUNCTION

A primary use for an unloading valve is associated with a dual pump circuit. Figure 10.11a illustrates an internally piloted (automatic control) circuit. Deliveries of both pumps are discharged into the circuit until the pressure approaches the setting of the unloading valve.

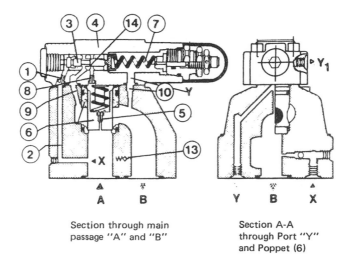

Section through main
passage "A" and "B"

Section A-A
through Port "Y"
and Poppet (6)

Figure 10.8 Multipurpose pilot operated valve structured for counterbalance service (internal pilot, internal drain). The counterbalance valve is used to keep the piston of a vertically mounted cylinder from falling freely because of gravity. Port A is connected to the lower cylinder port. This valve is an internally piloted valve. Pressure in Port A pressurizes spool (3) through the pilot line X. At the same time, pressure is present through the orifice (5) in the spring chamber, balancing poppet (6) closed. When pressure exceeds the setting of the spring (7), the pilot spool (3) moves allowing oil from the poppet (6) spring chamber to be relieved through orifice (8) and passages (9, 14) respectively into Port B. The poppet (6) lifts from its seat and allows the main stream of oil to flow from Port A to Port B. As long as the pressure in the pilot line X exceeds the setting of the spring (7), the spool (3) will remain open to maintain the oil flow from A to B. Port B is connected to the A or B port of the directional control valve. The valve should be set slightly higher than the maximum load induced pressure and is recommended for applications where the load is relatively constant. This valve is available with or without a check valve (13) between ports A and B. Courtesy of Rexroth Corporation.

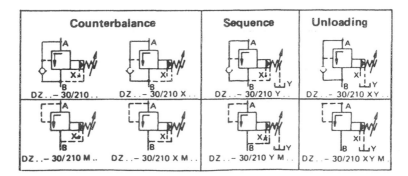

Figure 10.9 Symbols illustrating connections for counterbalance, sequence, and unloading function for the valve of Figure 10.8. Courtesy of Rexroth Corporation.

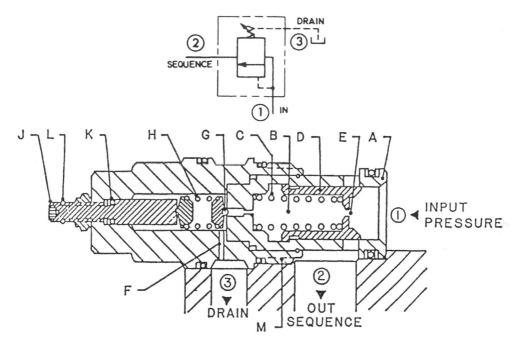

Figure 10.10 Piloted sequence valve in cartridge format. (A) Floating body, (B) control chamber, (C) minimum pressure bias spring, (D) main flow closure, (E) pilot supply orifice, (F) pilot return to tank passage, (G) pilot closure, (H) pilot spring, (J) adjusting screw, (K) seal for adjusting screw, (L) overpressure limiter, (M) thread and static seal assembly. Courtesy of Sun Hydraulics Corporation.

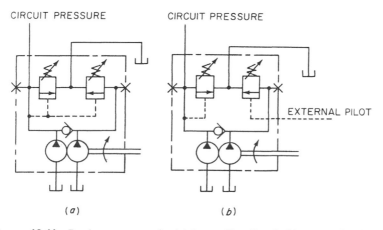

Figure 10.11 Dual pump controls. (a) Internally piloted, (b) externally piloted.

The large pump then discharges to tank at minimum pressure while the small pump continues to furnish oil to the circuit until the pressure setting of the integral relief valve is reached. If the small volume, high-pressure pump is of the pressure-compensated variable-displacement type, the flow will decrease to that necessary to make up for slippage. The relief valve will then serve as a safety valve set at approximately 25% more than the pressure-compensation control within the high-pressure pump. Figure 10.11b illustrates an externally piloted (remote control) circuit whose function is similar to (a), except that the unloading valve is controlled by an external pilot. Thus, the unloading pressure can be remotely controlled, permitting operation at pressures above the setting of the unloading valve up to the pressure setting of the relief valve with both pumps as long as the external pilot pressure source is programmed accordingly.

DUAL PUMP UNLOADING MALFUNCTIONS

A major malfunction of automatic circuit is improper adjustment of unloading valve. The pressure of unloading valve is typically fifty percent or less of the relief valve or compensator setting. A pressure setting approaching that of the relief valve or compensator adjustment permits the high volume pump to pass flow through the check valve and into the high pressure circuit with resulting heat and excessive power use. The relief valve area will usually be very warm if both pumps are passing oil through the valve.

UNLOADING AN ACCUMULATOR

Accumulator circuits can store pressurized oil within a predetermined pressure range. Typical intermittent service requires that the accumulator(s) be charged automatically by a pump or pumps that can be idle during the period when the accumulator is charged. When the pressure level in the accumulator is at a set low value the unloading piston is relaxed and the pilot poppet closes (Figure 10.12a). The relief function is negated and pump flow passes through the isolating check valve into the accumulator circuit. At the desired maximum charge pressure the unloading piston urges the relief pilot piston to the open position with a snap-action (Figure 10.12b). The pump flow is diverted to tank at a low pressure value established by the bias spring in the main relief valve, and the isolation check blocks the flow from the accumulator circuit back into the unloaded relief valve.

Troubleshooting an accumulator unloading valve may involve the accumulator as well as the valve. See Chapter 15 for specific circuit checks. The relief and unloading functions can be checked using the relief valve troubleshooting data.

THE COUNTERBALANCE FUNCTION

The basic structure of a counterbalance valve is shown in Figure 10.13. Note the check poppet that allows free return flow. The difference in area between the seal area of the relief poppet and the piston with the seal provides a safety valve function. Pilot assist oil is supplied from the actuator line opposite to the line in which the valve is installed. A counterbalance valve may be required in each actuator line (Figure 10.14).

The counterbalance valve provides maximum safety if it is manifolded to the actuator

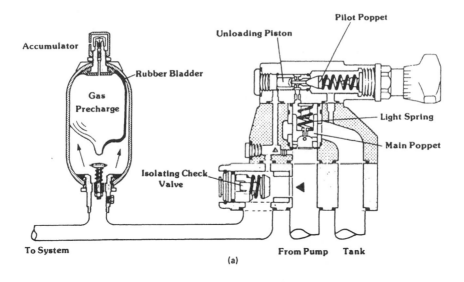

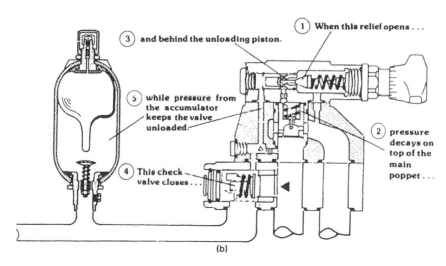

Figure 10.12 Accumulator unloading valve. (a) Charging the accumulator, (b) relieving and unloading of pump. Courtesy of Rexroth Corporation.

without piping or inserted in the actuator assembly in a cartridge format. Figure 10.15 illustrates a cartridge-type counterbalance valve with pilot assist at the nose of the valve.

The counterbalance valve of Figure 10.16 provides a convenient method of releasing the captive fluid in the event of a machine or power failure and the need to release a load. Check poppet "B" normally allows a free flow of oil from the directional valve (2) to the cylinder or motor (1). Spring "D" exerts a relatively low closing force for the check poppet. The check poppet can be opened to release the load gradually by turning the adjusting screw clockwise. Turning the adjusting screw counterclockwise increases the load holding pressure by compressing spring "E." Pilot flow at (3) is effective against the shoulder area of sleeve "C." Pilot pressure moves sleeve "C" until the stem of check piston "B" abuts the

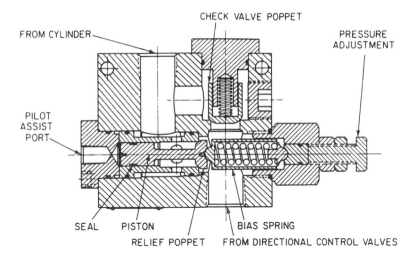

Figure 10.13   Counterbalance-type valve with return check and maximum pressure relief function.

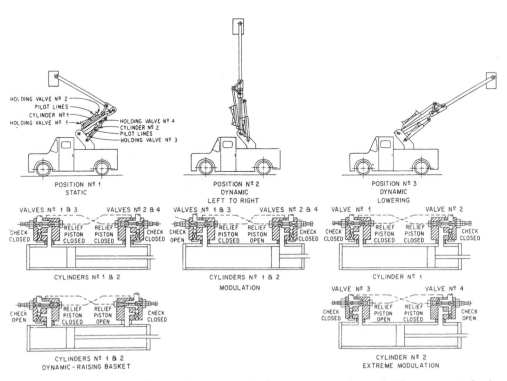

Figure 10.14   Counterbalance valve at each end of an actuator may be needed for over center loads.

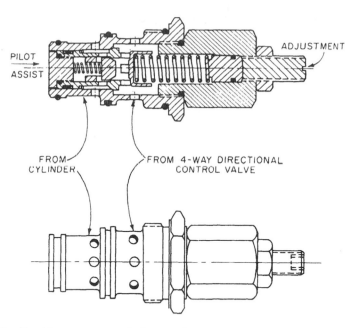

Figure 10.15 Cartridge-type counterbalance valve with pilot assist provided at the valve face.

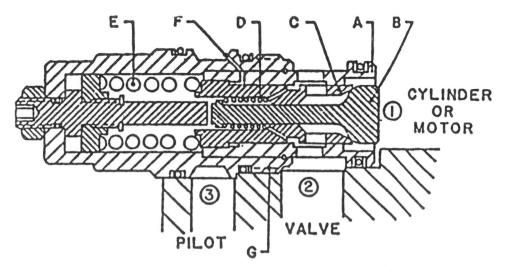

Figure 10.16 Counterbalance valve with safety release to lower personnel in bucket trucks or to gradually release a load. Courtesy of Sun Hydraulics Corporation.

adjustment extension at which time the trapped fluid can pass from (1) to (2). Pilot ratios of varying values can be provided.

COUNTERBALANCE VALVE MALFUNCTIONS

Load is erratic and valve chatters:

- The pilot ratio may be incorrect. Review the installation with sales engineer and make change to appropriate pilot ratio.
- The damping orifice may be needed in pilot line. Seek recommendation from the manufacturer.
- The ratio of 3:1 or 4.5:1 may provide better response characteristics particularly with swinging or winch loads.

Load drifts:

- Check for a leak in actuator.
- Inspect for contamination in the oil that may be holding valve off the seat.
- Inspect for wear resulting from contaminated oil.

Valve overheats:

- Check the pilot ratio. Traction drive may function best with 10:1 ratio holding the valve completely open. A lower ratio may cause heat.
- The valve set is higher than necessary to hold the maximum anticipated load.

11
Pressure-Reducing Valves

Pressure-reducing valves (PRVs) maintain secondary, lower pressures in branches of hydraulic systems. (The upstream main system pressure is still determined by the system relief valve or other pressure-setting device.) Pressure-reducing valves are normally open 2-way valves that allow system pressure fluid to flow through them until a set pressure is reached downstream. They then shift to throttle flow into the branch.

Pressure-reducing valves are actuated by forces exerted by pressure downstream. These forces establish the desired working pressure by creating a pressure drop across the valve's spring-biased main spool. A PRV is not an on-off device: the position of its main spool adjusts continuously to maintain the desired pressure setting. Of all pressure-control valves, PRVs are most sensitive to contamination-related malfunctions.

A PRV can malfunction in a variety of ways. To troubleshoot a PRV, install pressure gauges to read inlet pressure at test port TP1 and outlet pressure at test port TP2 (Figure 11.1), and use these readings to examine:

Decaying set pressure (low pressure at port TP2). If pressure at the outlet port drops below the desired set pressure, check the pilot head cone and seat for excessive wear which may permit increased drain flow (Figure 11.2). Excessive drain flow through this section of the valve reduces the pressure needed in the chamber above the main spool to increase valve pressure drop and bring operating pressure in the branch circuit up to the desired value.

Valve will not hold a reduced-pressure setting (high pressure reading at port TP2 of Figure 11.1). If preset pressure exceeds desired values, check for:

- A plugged pilot drain line; this would increase pressure in the chamber above the main spool, allowing the main system pressure fluid to flow into the branch circuit.
- A main spool stuck in the open position because of contaminants wedged between the spool and its bore, or because of scoring of the main spool or bore or both.

Valve cannot be adjusted to the desired low-pressure setting (high pressure reading of port TP2). If the valve cannot be adjusted to its desired pressure setting after the adjusting knob has been turned to its closed or nearly closed position, check for:

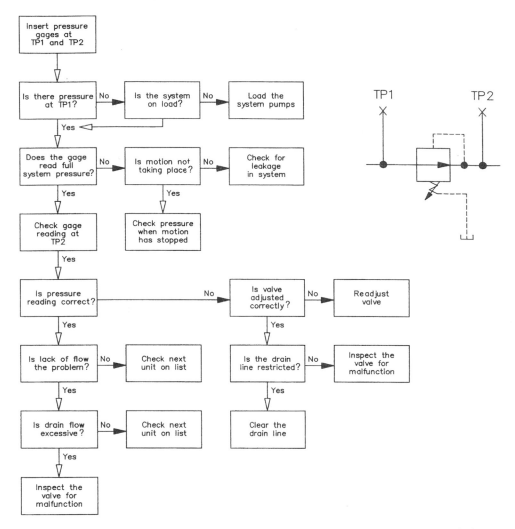

Figure 11.1 Steps to follow for troubleshooting pressure-reducing valves. Courtesy of Vickers, Incorporated.

- spool or bore wear, allowing main system pressure fluid to flow to the branch circuit.
- a broken spring in pilot head, resulting in inadequate cone-to-seat force in the control head.

Fluctuating pressure or no pressure at output port (zero pressure reading at port TP2). If there appears to be no fluid pressure at port TP2, check to establish if:

- The main spool is stuck closed, allowing no pressure fluid to flow to the branch. This condition can be caused by contaminants plugging the orifice in the passage connecting the two ends of the main spool, upsetting the hydraulic balance between the pilot pressure in the lower and upper control chambers (Figure 11.2 or 11.3), or
- the main spool is stuck closed because of contamination or scoring of the spool and/or its bore.

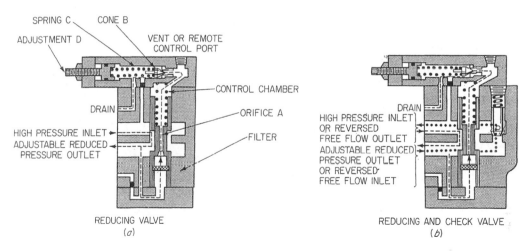

Figure 11.2 Piloted pressure-reducing valve. (a) For normal flow in one direction only or with limited return flow. (b) Equipped with integral check for free return flow through the valve. Filter is installed prior to orifice to minimize potential for plugging by contaminants in the oil.

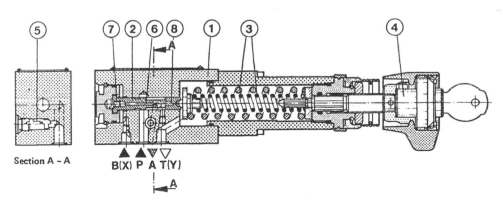

Figure 11.3 Direct operated subplate mounted pressure reducing valve. Housing may be a subplate mounted as shown or as a sandwich assembly to be installed between other valves and/or subplates. Valve consists of a housing (1), control spool (2), springs (3), and pressure adjustment (4) shown with optional locking knob to avoid unauthorized change of pressure level, an optional check (5) is available. The pressure adjustment (4) controls the pressure in the secondary circuit. The valve is normally open (i.e., oil may flow from port P to port A). A port pressure is sensed through pilot passage (6) and orifice (7) on the end of the spool, opposite the spring (3). When the pressure in port A increases to the value set by spring (3), the control spool (2) shifts into the control position and holds the set value constant in port A. Both the signal and control oil pass internally via orifice (7) and pilot passage (6) from port A. If the pressure in port A should rise above the set pressure, due to external forces in the secondary circuit, the control spool (2) moves further against the spring (3), thus allowing port A to connect via notches (8) in the control spool (2) to tank. Oil then flows back to tank to prevent any further rise in pressure. Pilot oil return from the spring chamber is drained external at port Y (tank port). An external pilot, if required, may be supplied at port X. The check valve (5) is required in order to allow free flow from port A to port P. Courtesy of Rexroth Corporation.

PILOT PRESSURE FROM THE PRIMARY SOURCE

Precise pressure settings can be maintained, even at high flows because the piloting function is supplied from the primary, high pressure inlet, rather than the secondary. This prevents turbulent flow, downstream of the regulating orifice, from influencing the quantity of pilot oil supplied.

In reference to the cross-section view (Figure 11.4) the valve is normally open, allowing oil flow from the primary (Port B) to the secondary (Port A) through the main spool (1). Up to the pressure setting of the pilot section (Figure 11.5), system pressure is maintained on the top (spring-loaded side) of the main spool (1) by feeding through pilot passage (2), pressure-compensated flow control (3), and orifice (4). This maintains the main spool in an open position.

When pressure in the pilot chamber exceeds the setting of spring (5) the pilot poppet (6) opens. The pressure-compensated flow control maintains a constant pilot oil flow while the pilot relief valve limits the maximum pressure holding the main spool (1) in the open position. When pressure in the secondary port (Port A) exceeds the pilot pressure plus spring force, the main spool (1) moves into a modulating control position, thus maintaining the secondary pressure at the valve's setting. Since the main orifice is created by alignment of radial holes in the main spool and sleeve, critical clearances are not subjected to high velocity oil flows, as they are in spool-type reducing valves. This provides smooth response and long life.

In the static, no-flow condition of the valve (main spool fully closed), an overload at the actuator or leakage oil tends to increase pressure in the secondary port (Port A). Relief protection is provided through the pilot section by a miniature relief valve poppet (7) built into the main control spool (1). The pilot section (10) must always be externally drained through the Y port (8 or 9).

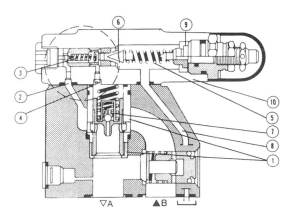

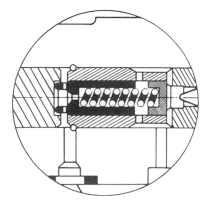

Figure 11.4 Piloted reducing valve with pilot pressure from primary source. Provides a uniform pilot pressure flow with minimum variations which might occur from varying turbulent flows in the secondary port. Courtesy of Rexroth Corporation.

Figure 11.5 Compensation for primary source pilot pressure. Orifice and bias spring limit rate of flow into the control area for proper functional control action. Courtesy of Rexroth Corporation.

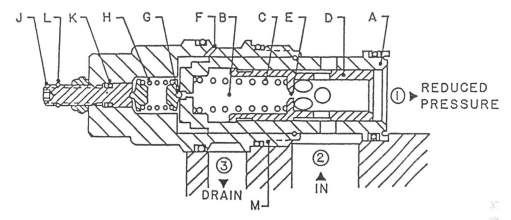

Figure 11.6 Cartridge-type reducing valve. Drain must be connected to tank with minimum resistance. Round holes in major closure assembly provides smooth passage with minimum transitional shock. Back pressure will be additive to closure (G) and spring (H). (A) Floating body, (B) control chamber, (C) minimum pressure bias spring, (D) main flow closure, (E) pilot supply orifice, (F) pilot return to tank passage, (G) pilot closure, (H) pilot spring, (J) adjusting screw, (K) seal for adjusting screw, (L) overpressure limiter, (M) thread and static seal assembly. Courtesy of Sun Hydraulics Corporation.

PILOTED REDUCING VALVE IN CARTRIDGE FORMAT

The cartridge format permits installation of the PRV in manifold assemblies, actuator housings, and/or in specialized housings (Figure 11.6). The drain port must return to the reservoir without restriction. Back pressure will be additive to spring "H" and pilot closure "G."

PILOTED REDUCING VALVE WITH AN OVERPRESSURE RELIEF

The closure assembly can be supplemented with a tank port that will pass a major flow in the event of pressure build up in the secondary port above set values because of external forces. The symbol for the piloted reducing valve and overpressure relief valve is shown in Figure 11.7. The pilot drain passes into the major tank passage (Figure 11.8). Cartridge valves can normally be removed, cleaned by ultrasonic vibration systems, and returned to service without altering pressure level adjustment.

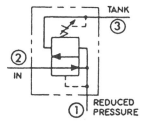

Figure 11.7 Symbol for pressure-reducing valve with integral overload relief valve.

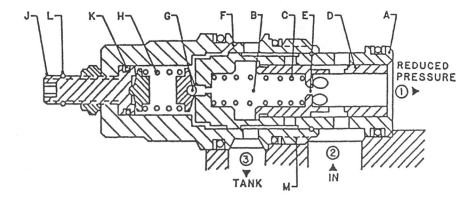

Figure 11.8 Cartridge-type reducing valve with integral overload relief valve. Pilot drain oil is returned to major tank connection. Tank line must pass freely to the reservoir to avoid back pressure which is additive to spring (H) and closure (G). (A) Floating body, (B) control chamber, (C) major flow bias spring, (D) main flow closure, (E) pilot supply orifice, (F) pilot return to tank passage, (G) pilot closure, (H) pilot spring, (J) adjusting screw, (K) seal for adjusting screw, (L) overpressure limiter, (M) thread and static seal assembly. Courtesy of Sun Hydraulics Corporation.

12

System Tests for Flow Control Valves

Frequently, as actuators cycle, loads vary, creating variable pressure drops through the fluid conducting lines. This often results in undesirable actuator speed variations. An effective means of insuring a uniform actuator rate is the use of pressure-compensated flow control valves. These restrictive devices maintain a constant fluid flow through a fixed or variable orifice in the valve. In response to workload changes, the valve's compensator spool moves toward an open or closed position. Thus a pressure-compensated flow control valve incorporates a device to ensure a constant pressure drop over the valve's variable or fixed orifice or *throttle*. The established fixed orifice or the setting of the variable orifice determines flow through the valve. The compensator controls the pressure at the fixed or variable orifice, resulting in smooth, constant flow through the valve regardless of load changes resulting from the load encountered by the actuator within the pressure and flow ratings of the valve.

FIXED ORIFICE FLOW RATE REGULATORS

Note the combination fixed orifice and modulating spool of Figure 12.1a. Fluid entering the inlet port meets the resistance created by the face of the modulating spool and the relatively small hole in the face of this restrictive device. The modulating spool is pushed against the bias spring; this tends to limit the flow at the cutoff point. Fluid flows through the fixed orifice balancing the pressure on each surface of the modulating spool. This causes movement of the modulating spool because of the balanced areas and the force of the bias spring. The resulting movement minimizes the restriction at the cutoff point in the path to the outlet. The flow through the orifice will equal the flow through the cutoff point. Thus this restriction will pass a constant flow that is determined by the size of the orifice and the installed value of the bias spring. Flow can be increased by enlarging the fixed orifice and/ or increasing the spring force. Flow can be decreased by reducing the size of the fixed orifice and/or reducing the force exerted by the bias spring.

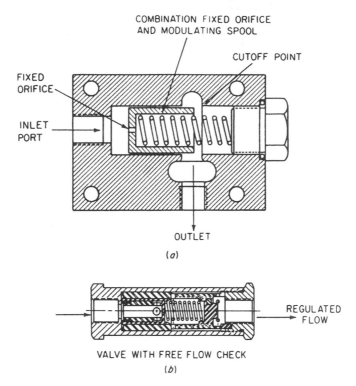

Figure 12.1 Nonadjustable compensated flow control valve. (a) Compensated flow rate is increased by larger orifice size or heavier spring. Flow rate is decreased by smaller orifice or lighter spring. (b) A free flow check can be included in the design.

INTEGRAL CHECK FOR FREE RETURN FLOW

The cross-section view of the valve of Figure 12.1b shows the introduction of an integral check valve for free flow in the reverse direction. Valves with the general configuration of Figure 12.1b have been used for many years to establish the rate at which the forks of a lift truck descend, regardless of the magnitude of the load. The rate of descent is controlled by this nonadjustable valve at an appropriate speed that permits safe operation as the machine is operated with personnel having various operating skill levels. The integral check valve permits raising the load at a rate that can be established by the operator manipulating the directional valve.

COMPENSATED ORIFICE OPTIONS

A pressure-compensated flow control valve can combine a pressure-reducing valve and needle valve in the same housing (Figure 12.2). The orifice may be created by a spigot-type mechanism with a limited rotation to create the desired reduced passage appropriate to the application needs. This is an option that permits the use of a dial where settings can be identified. For many applications this provides needed reference points for machine rate-of-

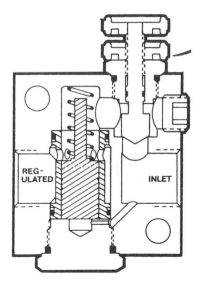

Figure 12.2 Adjustable compensated flow control valve with needle valve orifice and restrictive-type compensation. Pressure drop will be a function of the installed value of the bias spring and the area of the regulator spool.

Figure 12.3 Spigot-type adjustable orifice for pressure-compensated flow control valve. (a) Sectional view. (b) Schematic showing pilot connection at the inlet to the orifice and pilot connection at the outlet of the orifices sensing the downstream load.

travel adjustments (Figure 12.3). The bias spring cavity of the reducing spool assembly of Figure 12.3a is ported to the load port at the outlet from the spigot instead of being ported to atmosphere. The pressure level of the load created by the actuator is then added to the value of the bias spring, and pushes the reducing spool to the open position. The load-sensing line of the pressure-reducing element that pushes the spool against the bias spring to reduce flow through the compensator is connected to line pressure in the cavity between the outlet of the reducing spool and inlet to the spigot (or other control orifice as designed for the specific needs). Thus the pressure drop across the orifice (spigot or other restrictive element) is constant regardless of input to the reducing spool or load pressures at the outlet created by the actuator within the design characteristics of the flow control valve. With this constant pressure-drop value across the imposed orifice the flow rate through the valve is constant.

Two important features are incorporated in the pressure-compensated valve structure of Figure 12.4. The first is a locking dial that can prevent undesired changes of the valve setting. The second feature is an adjusting screw for fine tuning the maximum size of the passage through the compensator. This adjustment prevents an undesired flow surge while the compensator is automatically positioning itself to provide the desired flow across the adjustable orifice. The compensator is set initially for a desired maximum flow through the valve. This maximum passage can be adjusted to the specific flow needs as established by the machining process and may be significantly less than the maximum designed flow capability. Thus a machine that has a momentary erratic movement as the flow control valve begins its function can be corrected by fine tuning the functional positioning of the compensator spool.

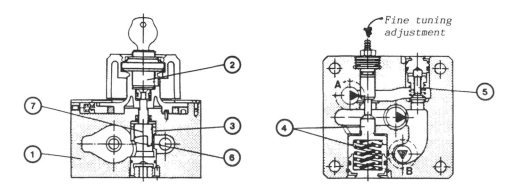

Figure 12.4 Spigot-type orifice with lockable knob. The valve includes a housing (1), lockable adjustable (2), orifice (3), pressure compensator (4), fine tuning adjustment, and check valve (5). At position (6) the flow is throttled from A to B. The throttle section is adjusted by turning the spiral-shaped orifice spool (7). A pressure compensator (4) is used to maintain constant flow independent of pressure at the throttle (6). Flow variations is not affected by changes in temperature and viscosity due to the sharp-edge orifice principle. There is a free flow from B to A via the check valve (5). Courtesy of Rexroth Corporation.

The valve of Figure 12.5 has also been modified to fine tune the compensator by use of a special screw jack or suitable nonadjustable stop. The pressure drop across the compensator is also adjustable by changing the installed bias spring value with the adjusting screw provided in the rear cap.

The micrometer-type adjustment for the triangular orifice of Figure 12.6 can be replaced by a proportional-type direct-current solenoid. The adjustment is by linear thrust for positioning of the orifice mechanism to create the desired size. A simplified proportional

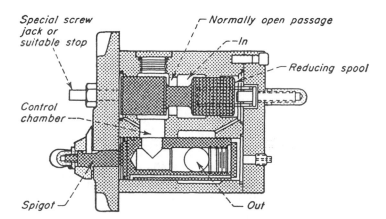

Figure 12.5 Pressure drop across the adjustable orifice can be increased or decreased by changing the installed length of the compensator bias spring with a threaded adjustment screw. The maximum flow through the reducing spool to the control chamber can be adjusted with the special screw jack or an inserted stop slug.

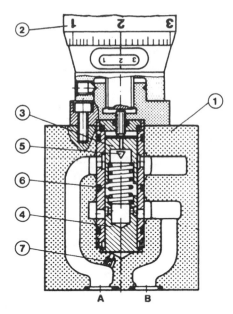

Figure 12.6   Flow control with linear micronic adjustment. Valve consists of housing (1), adjusting element (rotary knob) (2), orifice (3), pressure compensator (4), and, if required, a check valve. At throttle position (5), the oil flow is throttled from port A to port B. The opening of the orifice is adjusted by rotary knob (2). In order to hold the fluid flow constant and independent of the pressure in port B, a pressure compensator (4) is set after the throttle (5). A spring (6) holds the pressure compensator (4) in the open position when no oil is flowing through the valve. As soon as fluid flows through the valve, pressure from port A passes through orifice (7) and generates a force on the pressure compensator (4). The pressure compensator (4) moves until the forces on it are balanced and it achieves a regulating position. If the pressure rises in port A, the pressure compensator moves towards a closed position, until the forces are balanced. Because of the constant "following action" of the pressure compensator, a constant flow rate is achieved. Courtesy of Rexroth Corporation.

flow control valve symbol is shown in Figure 12.7. A complete symbol for proportional flow control valve is shown in Figure 12.8.

The proportional flow control valve of Figure 12.9 accepts a preset electrical signal. This signal causes the valve to control flow independent of pressure and temperature variations. The assembly consists of a housing (1), a proportional solenoid with inductive positional transducer (2), measuring orifice (3), pressure compensator (4), together with stroke limiter (5), and optional reverse flow check (6).

The desired oil flow is set using a potentiometer, which in turn gives an input value to an amplifier. The amplifier, in turn, controls the orifice via the proportional solenoid. The position of the spool creating the orifice is determined by the inductive positional transducer, and any difference from the given input value is then corrected by the control system. The pressure drop across the measuring orifice is held constant by the pressure compensator. The oil flow is therefore independent of load. The low temperature drift is minimized by a sharp edge orifice design. With an input level of zero, the measuring orifice is closed. In the event of electrical failure, the measuring orifice closes. A smooth start is possible from the input value 0. The orifice may be opened and closed gradually by two ramps in the electronic amplifier. Free flow from B to A is possible via the check valve (6). Hydraulic

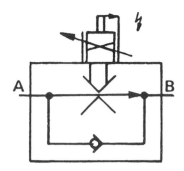

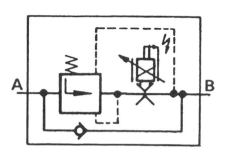

Figure 12.7 Simplified proportional flow control symbol with sharp edged orifice. Courtesy of Rexroth Corporation.

Figure 12.8 Complete symbol for proportional flow control valve. Courtesy of Rexroth Corporation.

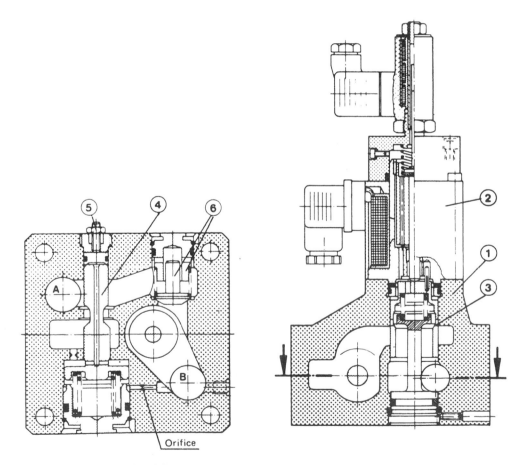

Figure 12.9 Proportional flow control valve. Rate of compensated flow established by means of a direct-current solenoid. Orifice size is monitored by a linear variable differential transformer feedback signal. Courtesy of Rexroth Corporation.

troubleshooting involves procedures outlined later in this chapter. Electronic troubleshooting is not addressed in this volume.

There are two types of pressure-compensated flow control valves: bypass and restrictive. The restrictive-type is most commonly used for machine tools because it can be installed in the circuit for meter-in, meter-out, and bleed-off functions. Bypass flow controls will be discussed later in the chapter.

System circuit designers must make sure that there is a pressure differential between system pressure and actuator working pressure. The designer can achieve this by sizing the actuator so that load pressure is lower than system pressure. The level of pressure drop required to maintain good compensator response varies with valve designs. Typical pressure drops range between 75 and 175 psi. Erratic actuator movement or changes in set speed indicate improper valve performance or inadequate pressure differential.

Troubleshooting flow control valves can be made easier by installing pressure gauges at inlet port TP1, outlet port TP3, and internal pressure chamber TP2 (Figure 12.10).

Typical problems encountered and remedial actions are:

Inadequate pressure differential between main system pressure and actuator working pressure   Check for adequate pressure differential using gauges at TP1 and TP3. Follow the valve manufacturer's recommendations.

Sticking compensator spool   Check pressure differential between the inner pressure chamber and valve outlet pressure, using gauges at TP3 and TP2, respectively. The pressure differential should approximate the compensator spool spring load. Fluid contamination and/or spool body scoring can cause erratic spool movement or hang-up. Either can affect the constant pressure drop needed across the throttle orifice.

Worn or contaminated return flow check valve   Problems with free-flow return check valves can contribute to faulty operation. Inspect for worn seat/poppet or presence of contaminants; either can allow oil flow through the check valve in the direction of controlled flow. An open check valve can cause identical pressure gauge readings at TP1 and TP2.

Drain line in flow control valve not vented to atmosphere   This is needed in valves requiring draining at the top of the throttle to get rid of internal leakage. If the drain port is not ported to atmosphere, the valve may rupture in the area of the adjusting handle.

In summary, fluid conditioning, worn internal parts, and the absence of good circuit design will cause pressure-compensated flow control valves to malfunction. Circuit considerations may need to be reviewed. Choice of meter-in, meter-out, or bleed-off can affect the service life and performance of specific tasks.

METER-OUT CIRCUITS

The compensated flow control valve used for precision machine feeds functions best with a relatively large flow of oil. Note the position of the flow control valve in Figure 12.11. Oil is metered out of the head end of the cylinder to obtain maximum flow through the valve. Metering from the rod end can result in pressure intensification and accelerated wear on the packing gland.

The meter-out circuit (Figure 12.12) is found on reciprocating machinery such as a

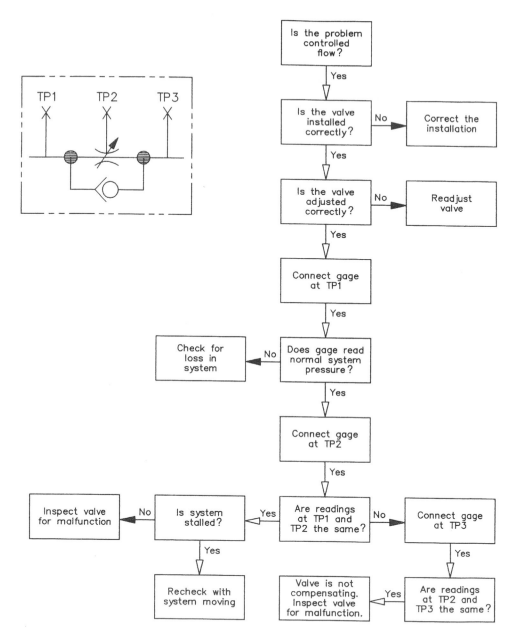

Figure 12.10 Steps for troubleshooting pressure-compensated flow control valves. Courtesy of Vickers, Incorporated.

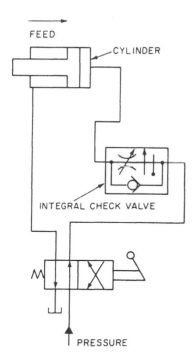

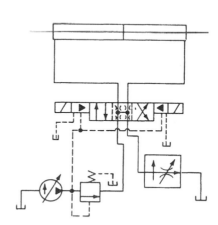

Figure 12.11 Meter-out circuit for precision machine tool feeds.

Figure 12.12 Meter-out circuit for reciprocating machinery.

grinder. A sequence valve is inserted to insure pilot pressure for the directional valve. The pressure drop needed may also be provided by an inline check valve with a spring of suitable value to insure the minimum pilot pressure needed to pilot the valve. The double-rod cylinder will not create an intensification condition.

METER-IN CIRCUITS

Forming presses, molding presses, and comparable high force applications function efficiently with a meter-in circuit (e.g., Figure 12.13). Many special material handling functions work well with a meter-in circuit. As an example, the circuit of Figure 12.14 was developed for the controlled movement of a basket associated with a quenching operation in a heat-treating facility. The cylinder is used on a hoist to lower a basket of parts into a quench tank adjacent to the heat-treating furnace. The cylinder must provide the power to lift the basket of white-hot parts gently out of the furnace without damaging the furnace walls. The basket is then swung over the quench tank and dropped quickly into the quench oil. As the valve is shifted to let the cylinder drop the load, the fluid enters the head end of the cylinder without restriction. The pressure created by the load on the rod end of the cylinder forces the fluid through the integral check valve within the flow control valve assembly. Cavitation on the head end of the cylinder is eliminated by check valve X that adds capacity to the pump to prefill the cylinder during the rapid drop.

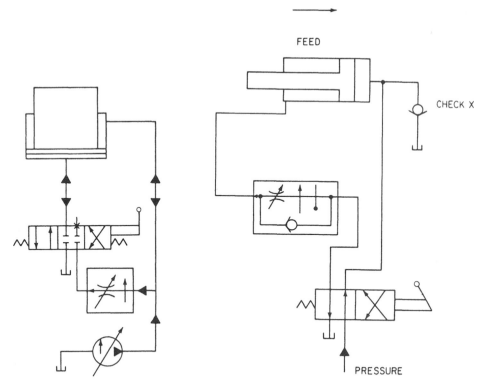

Figure 12.13 Meter-in circuit for forming, molding, and other high force machines.

Figure 12.14 Meter-in lift-circuit with check valve for prefill service.

BLEEDOFF CIRCUITS

Line pressure in the bleedoff circuit of Figure 12.15 will normally be limited to that needed to move the load. A safety relief valve is included in the event the cylinder is not stopped prior to reaching the end of the stroke. Bleedoff circuits have been widely used for reciprocating tables, broaching, billet gougers, and comparable machining functions.

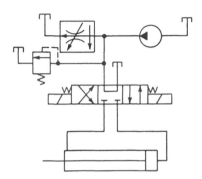

Figure 12.15 Bleedoff circuit with emergency safety-type relief protection.

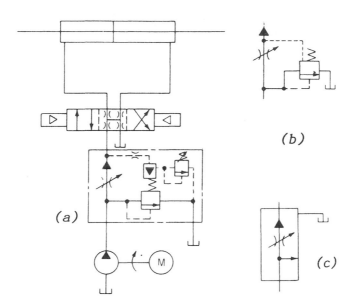

Figure 12.16 Bypass-type flow control circuit. (a) Complete circuit illustrated schematically. (b) Circuit illustrating the position of the relief valve and load-sensing pilot porting. (c) Simplified symbol.

BYPASS-TYPE FLOW CONTROLS

Note the symbol of Figure 12.16a. The enclosure indicates that the components are incorporated in one assembly. In this instance it is the adjustable orifice, a piloted relief valve connected ahead of the orifice, and a pilot line from downstream of the orifice to the control chamber of the relief valve. A pilot is also included to establish a maximum safe operating pressure. The pressure drop across the orifice is established by the value of the bias spring within the piloted portion of the relief valve. The load pressure level reflects to the top of the piloted relief valve closure. The pilot line to the bottom of the relief valve closure is shown in the symbol. The piloted closure has equal areas on top and bottom and the bias spring creates the pressure value to direct the flow through the adjustable orifice. Pressure in the circuit is that of the reflected load plus the piloted relief valve closure bias spring.

A bypass-type flow control valve can provide an efficient functional circuit with the limitation of only one control per circuit. This flow control valve cannot be used in parallel in the same manner as the restrictive-type flow control valves.

Figure 12.16b illustrates schematically the operation of the bypass flow control compensation. Figure 12.16c is a simplified symbol.

13

Directional Control Valves

This chapter discusses major types of directional control valves. Included are those operated manually and those with direct electrical operation, cam operation, and direct solenoid operated pilot valves, and the piloted directional valves that they control. Low flow capacity directional valves actuated by proportional direct current solenoids will be discussed as well as proportional and electrohydraulic servo pilot valves and the piloted directional valves that they control. Included in the materials will be the alternating current pulse-width modulated pilots used for positioning the directional valve spool by means of a piloting system and linear bidirectional electric motors used for positioning the flow directing element.

Malfunction of a solenoid valve is often the result of maintenance deficiencies. The malfunction may cause erratic operation or complete solenoid failure. These maintenance deficiencies may include fluid contamination affecting valve performance and ultimately resulting in excessive internal leakage or jamming of the valve elements. There may be insufficient pilot pressure to shift the second stage valve spool in an appropriate pattern. An improper electrical signal to the solenoid can result in pilot malfunction or damaged solenoid coil. Excessive tank/drain line pressure in the pilot valve drain line can adversely affect the input pilot signal.

The initial troubleshooting techniques that follow are segregated into the two basic solenoid operated valve classifications: direct and pilot operated valves. Figure 13.1 illustrates a system test for directional control valves plus a directional valve symbol showing position of test gauges.

SOLENOID OPERATED VALVES (DIRECT)

Malfunction   Actuator does not move following valve energization (system loaded).

- If gauges TP1 and/or TP2 do not show working pressure, operate the valve with the manual override to shift pilot stage (Figure 13.2). If spool moves freely and actuator moves then the solenoid can be presumed to be bad or the electrical signal to the solenoid is not proper. Test electrical signal to the solenoid for proper voltage. If the

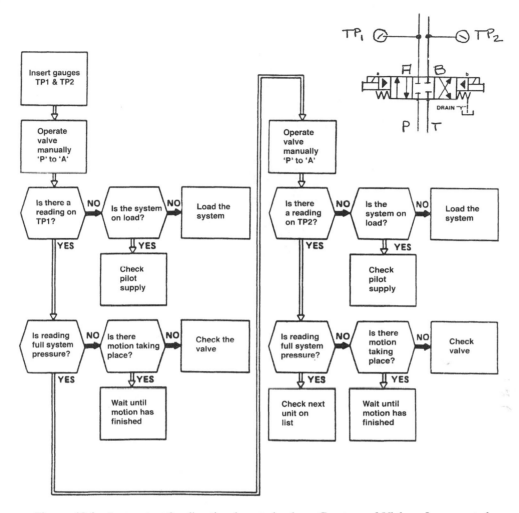

Figure 13.1 System test for directional control valves. Courtesy of Vickers, Incorporated.

voltage is proper the coil may have failed. However, the complete solenoid should be inspected to prove out operational worthiness.

- If spool does not move freely check spool and body bore for scoring, contamination, or other defects which impede free spool movement. Inspect O-ring seal on solenoid push pin if used. Replace if worn or excessively tight. Repair or replace the valve.

Note: Single-solenoid, spring-offset, two-position, spool-type valves when used in power failure safety circuits are susceptible to silting (build up of very small particles wedging the spool into a locked position) with attendant spool sticking. When the valve is held in an energized position for long periods of time internal leakage carries very fine contamination (silt) between the spool and its body bore. When the valve solenoid is deenergized following power failure the flow directing spool is expected to return to the spring offset (fail safe) position. The bias spring does not have sufficient energy to return the spool and solenoid armature because of the silting.

Correction Appropriate micronic filtration in supply lines to critical valves.

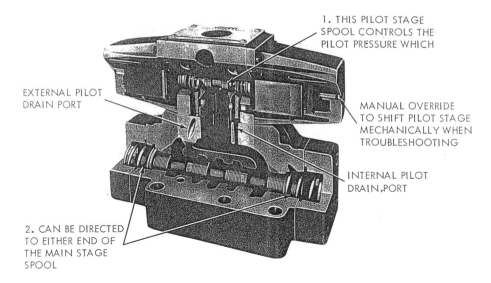

1. THIS PILOT STAGE
SPOOL CONTROLS THE
PILOT PRESSURE WHICH

EXTERNAL PILOT
DRAIN PORT

MANUAL OVERRIDE
TO SHIFT PILOT STAGE
MECHANICALLY WHEN
TROUBLESHOOTING

INTERNAL PILOT
DRAIN PORT

2. CAN BE DIRECTED
TO EITHER END OF
THE MAIN STAGE
SPOOL

Figure 13.2 Manual override can be used to shift the pilot valve spool in the troubleshooting procedures. Courtesy of Vickers, Incorporated.

SOLENOID-CONTROLLED PILOT-OPERATED VALVES

Malfunction   Actuator does not move following valve energization (system loaded).

- If gauges TP1 and/or TP2 do not show working pressure, manually operate the pilot valve solenoid push pin to check for improper or lack of electrical signal, solenoid failure, or pilot valve spool malfunction (Figure 13.2).
- If the pilot valve is operating properly then the problem is beyond the pilot valve (i.e., the second stage that is controlling system fluid flow). Failure of the second stage valve spool to react to pilot valve energization is the result of insufficient pilot valve pressure, contamination, or valve spool/body bore scoring. Other reasons also could be broken parts causing a mechanical lock up of the spool.

Correction for insufficient pilot pressure:

- The *external pilot pressure source* may be created by resistance valve in major line from pump (Figure 13.3) or resistance valve in the tank line from the piloted valve (Figure 13.4). The resistance valve for pilot pressure source may be in the body of the piloted valve (Figure 13.5). Insert gauge to be certain resistance valve is functional and the needed pilot pressure is available. The resistance valve may be inspected for foreign impediments in some models.
- The *internal pilot pressure source* may be created by a resistance valve in tank line from a piloted directional control valve (Figure 13.4). Insert gauge to be certain resistance valve is functional and the needed pilot pressure is available. The pilot pressure source may be taken internally from the piloted valve with a reducing valve sandwiched between the piloted valve and the pilot valve. Make certain the secondary pressure established by the reducing valve is set properly. Inspect the reducing valve spool and/or bore for sticking.

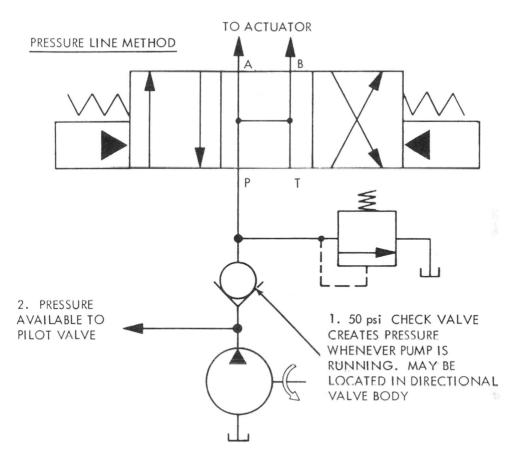

Figure 13.3 Pilot pressure sources can be created externally with a resistance valve on the supply line.

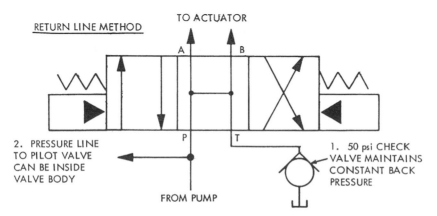

Figure 13.4 Pilot pressure sources can be created externally with a resistance valve in the return to tank line.

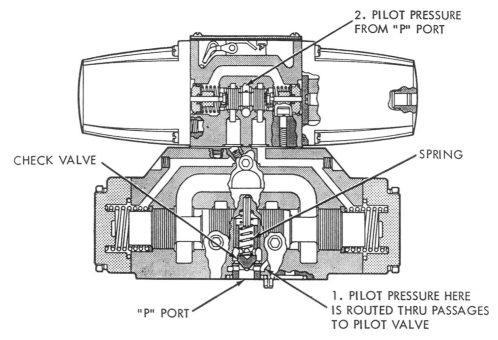

2. PILOT PRESSURE
FROM "P" PORT

CHECK VALVE

SPRING

"P" PORT

1. PILOT PRESSURE HERE
IS ROUTED THRU PASSAGES
TO PILOT VALVE

Figure 13.5 Resistance valve may be integrated in piloted valve body. Courtesy of Vickers, Incorporated.

Correction for pilot drain restrictions:

• Inspect external drain lines for restriction; clear, if necessary.
• The internal drain line may be affected by back pressure created during part of the work cycle; install external drain directly to tank.

Malfunction   Mechanical lockup from scoring of spool and/or bore.
Correction   Repair or replace components.
Malfunction   Inadvertent shifting of second stage valve spool (particularly critical in "no spring" detented valves).
Correction   Check for back pressure or surges in the pilot valve drain line by inserting a pressure gauge in the drain/tank line. Pilot pressure must always exceed tank or drain line pressure. Refer to manufacturers specifications for valve and spool-type minimum pilot valve pressures. Tank or drain line surges that reduce the required differential pressure can cause the second (main) stage spool to shift.
Malfunction   Decay in actuator speed of movement (assume the problem is with the valve).
Correction   Check for wear (clearance) between spool and body bore causing high internal leakage. In addition to generating heat, internal leakage (closed center valves) can cause actuator movement. Internal leakage flow through the A and B ports to a differential area cylinder can cause cylinder creep.
Malfunction   Actuator speed varies in the two directions of travel.
Correction   Inspect for stroke adjustment limiting maximum spool travel (Figure 13.6).

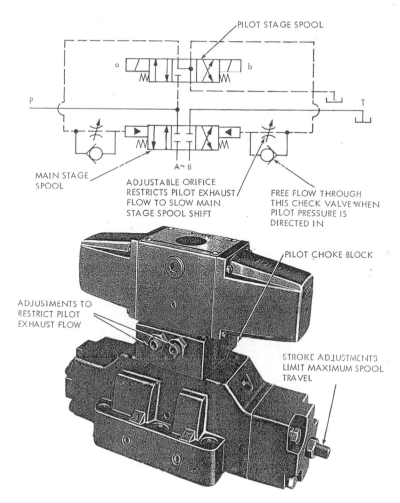

Figure 13.6 Pilot chokes and stroke adjustment limits can be incorporated in the valve assembly. Courtesy of Vickers, Incorporated.

Malfunction Piloted spool movement too slow (lost production) due to inadequate pilot pressure and/or the pilot choke mechanism excessively restricts pilot flow.
Correction

- Increase pilot pressure level.
- Adjust pilot choke mechanism for appropriate flow to provide the desired rate of directional valve spool movement.
- Replace digital solenoid for pilot valve with proportional solenoid and electronic control for pilot valve.

Malfunction Piloted spool movement too fast (shock). Internal pilot pressure too high making throttle adjustments too difficult.
Correction

- Supply pilot pressure via external pilot source through a small pressure reducing valve at pressures less than approximately 250 psi (Figure 13.7).

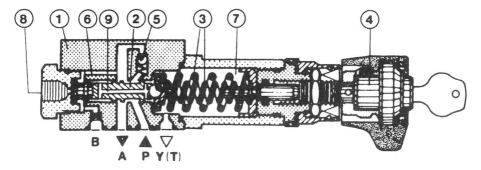

Figure 13.7 Small capacity reducing valve to supply pilot pressure to directional valve pilots. Housing (1), control spool (2), nested bias springs (3), pressure setting element (4), optional check valve (5) which is not needed for valves employed to direct pilot pressure to pilot directional control valves. Secondary (pilot pressure) is set by means of the pressure setting element (4). The valves are normally open (i.e., oil can flow freely from P to A). At the same time, pressure at A affects the spool surface opposite the springs (3) by means of control line (6). If the pressure in A exceeds the value set at the springs, the control moves the spool into regulating position and holds the value set in A constant. The signal and oil come internally from A by means of control passage (6). If, due to external forces at the user, pressure in A continues to rise, the control spool is pushed further against the springs. Thus A is connected to tank by means of an orifice in the control spool. Oil drains until pressure decreases. Pilot oil return from the spring chamber (7) is always exhausted to Y (tank port). Secondary pressure can be read at port (8). Courtesy of the Rexroth Corporation.

- Install a sandwich-type pressure reducing valve between the internal pilot supply from the piloted valve and the pilot valve (Figure 13.8).
- Replace digital solenoid valve with proportional solenoid valve using electronic control for rate of spool movement.

PILOT-OPERATED DIRECTIONAL VALVES

A spring centered pilot operated four-way is shown in the upper view of Figure 13.9. To insure positive centering the pressure centering mechanism shown in the lower view of

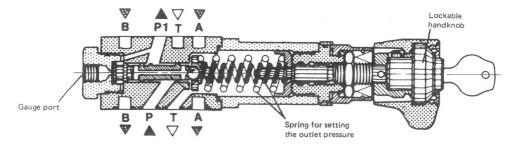

Figure 13.8 Reducing valve to sandwich between pilot valve and piloted valve. Reduced pressure (P1) can be checked at gauge port. A and B ports pass through the valve block without restriction. The tank port also passes through the block without restrictions and also drains the spring cavity of the reducing valve. Courtesy of Rexroth Corporation.

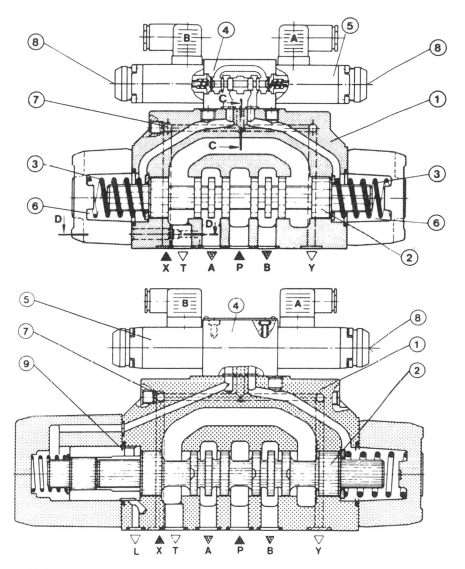

Figure 13.9 Solenoid controlled, pilot operated four-way valve: spring-centered (upper figure). Pressure-centered (lower figure). Courtesy of Rexroth Corporation.

Figure 13.9 can be employed. The spool (2) is fitted in housing (1) and centering springs (3) hold the spool in neutral with pilot pressure removed from the end cap cavities. The pilot valve (4) is operated by solenoids (5). Manual operators (8) provide actuation for set up purposes and for trouble shooting. Spring-centered models release pilot pressure from the cavity (6) in the rest position and connect the cavity to the pilot drain port. The pressure-centered model directs pressure to both end cavities in the neutral position to allow the pressure-centering assembly (9) full control over the centering function of the main spool (2). The cavity between the end of the main spool and the centering mechanism (9) must be drained via port L. Port X is the pilot pressure connection that can be arranged for either internal or external pilot source. Port Y is the pilot drain that can be either internal or external depending on the circuit requirements.

DIRECTIONAL CONTROL VALVE ASSEMBLIES FOR MOBILE MACHINERY

To simplify piping and conserve space, many mobile hydraulic systems use two or more spool-type directional valves to control independent functions in a single housing. Often the relief valve (item 1, Figure 13.10a) is included in the assembly. This valve (in cartridge form) is inserted in the directional control valve entry to provide protection at the point of potential shock generation without need for additional circuit piping. Usually one supply line and one return line to tank provides the needed supply and return functions. The cylinder lines are connected to the appropriate actuators within the assembly.

Models can be provided with facilities to permit power *beyond* the initial valve assembly. Two or more valves can operate in series with one pump supplying both valves. A return port from each valve is usually run independently to tank. Power beyond port from the upstream valve is connected to the pressure inlet of succeeding valves.

Valves specifically designed for mobile or marine service (e.g., Figure 13.10) often incorporate virtually all of the major control functions in a single assembly. For troubleshooting purposes it is essential that a circuit drawing is available to determine the functions provided in each section of the valve assembly.

CARTRIDGE VALVES ELIMINATE NEED FOR EXTERNAL PIPING

Essential functions may be provided in many components by inserted cartridge valves within the control valve assembly. One typical function of these inserted cartridge valves is the relief function at the entry. Relief valves can also be placed at the outlet of the spool prior to the connection to the cylinder port to protect the line if the spool is closed and external forces increase the pressure level in the cylinder lines. The relief valve tank connection can be connected to the opposite cylinder port or directly to tank as appropriate to the operating needs. Anticavitation check valves can also be located in this same cylinder port area in cartridge format to avoid cavitation in the actuator lines. The anticavitational check can be supplied from the integral tank porting within the directional valve casting.

CYLINDER CONNECTIONS MAY BE OBVIOUS BUT SPOOL DATA NEEDS SYMBOLS

The functional components of a lever-type assembly usually can be traced by following the cylinder lines to the actuator assembly. The internal functions must be identified by reference to the circuit drawing.

The valve assembly shown in Figure 13.10a has a supply line at position 2. The relief valve (1) is pilot operated and externally adjustable. The discharge port (3) can have facilities for series connection to another bank of valves (power beyond) with the capability to connect a return line for the upstream valve. In the neutral position, the inlet (2) is connected to the outlet (3) via the series connections through the directional spools and

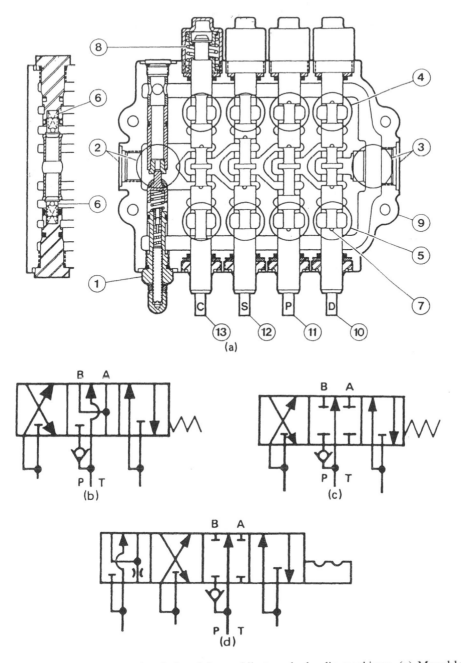

Figure 13.10 Directional valve designed for mobile type hydraulic machinery. (a) Monoblock sectional view, (b) motor spool symbol, (c) spool for double-acting cylinder, (d) four-position spool to include float position. Courtesy of Parker Hannifin Corporation Mobile Hydraulic Division, Cleveland, Ohio.

valve coring within the valve housing (9). Cylinder ports (4) and (5) are either blocked in neutral or connected to tank according to the spool design. Spool C (item 13, Figure 13.10a) is identified as a motor spool. The interconnecting lines are shown in symbolic format in Figure 13.10b. Both cylinder lines are connected together and to tank in the neutral or center position. Spool P (item 11, Figure 13.10a) and spool S (item 12, Figure 13.10a) are designed to be used with a single acting cylinder. Spool S and P are shown in schematic form in the symbol of Figure 13.10c. Spool D (item 10, Figure 13.10a shown symbolically in Figure 13.10c) is used for a double acting cylinder. A four-position spool (shown symbolically in Figure 13.10d) can be provided for operation equal to that of spool D (item 10, Figure 13.10a). In addition, a float position has been added.

Check valves (6) (Figure 13.10a and shown symbolically in Figure 13.10b) are inserted between the inlet pressure port and the supply to the cylinder to prevent actuator drop as the spool is shifted. These are known as *load checks.*

Metering notches (7) (Figure 13.10a) provide better control of the actuator speed. The centering spring (8) can be replaced with mechanical detents, switch actuating mechanisms, and/or automatic kick-out devices as required by the machine function. A spring-centering mechanism provides a *dead-man* control so that actuator will stop or coast, according to spool center configuration, as the operator releases the operating lever. A detent mechanism will hold position, such as taking up slack on a winch line, when operator chooses to operate other functions simultaneously without the need to hold the detented spool. The detent is shown symbolically in Figure 13.10d. No action will occur unless the lever is moved to release the spool from the detent mechanism. An automatic kick-out device can provide a safety or energy saving function by releasing the spool at a predetermined signal value. Crossover reliefs and make-up checks can be supplied internally to the ports of the valve leading to the actuator. Crossover reliefs protect actuators when fluid is locked in the cylinder lines by the major flow directing spool. Make-up checks supply fluid to appropriate actuator lines when the control spool isolates these ports. Fluid passing to the tank from a rotary motor drain as it coasts to a stop at the rate established by the crossover relief valve setting can be replaced by the flow through the make-up check valve. Similarly, fluid lost because of cylinder port relief valves (directing fluid to tank) is supplied to the opposite port of the cylinder by means of the make-up check valve.

The circuit as shown in Figure 13.11a provides a parallel supply and parallel return to tank. Therefore, two or more spools can be engaged simultaneously. A check valve in the supply line to each valve segment prevents loss of load as other valves encounter lighter loads. Pressurized fluid will seek the lowest resistance to flow. With series pressure and parallel return to tank (Figure 13.11b), the engaged spool closest to the valve inlet takes all pump flow.

The exterior view of a six-spool valve assembly in a common body is illustrated in Figure 13.12a. Note the externally adjusted relief valve at the inlet port. Typical pressure drop versus flow curves for this six-spool valve are shown in Figure 13.12b. The monoblock structure of Figure 13.10a limits flexibility when mid-inlet connections are desired for special circuits. Segmented valve structures usually have the front inlet module fitted with three ports: top inlet, side inlet, and side outlet. Unused ports are plugged. Mid-inlet modules are fitted with a top inlet port and provide a method to pipe output from a secondary pump source to a bank of valves. A high degree of flexibility of design is provided with the modular structure shown in Figure 13.13a. Practically all features are available in the modular design. The exterior view of Figure 13.13b shows the optional anticavitation checks located by the cylinder ports. These are used to reduce the chance of

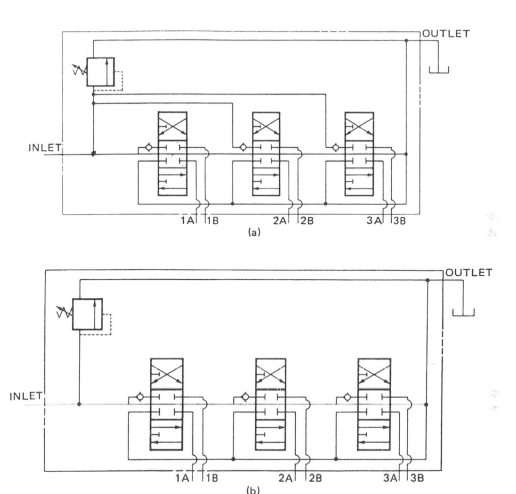

Figure 13.11   Parallel and series circuit. (a) Parallel supply and parallel return, two or more spools. (b) Series pressure, parallel tank line. The engaged spool closest to the valve inlet takes all pump flow. Courtesy of Parker Hannifin Corporation Mobile Hydraulic Division, Cleveland, Ohio.

cavitation due to overload of the circuit. The optional overload relief valves just below the check valves limit the pressure in the specific cylinder line and, of course, in the circuit when the pressure is connected to that port. Spool assemblies in mobile valve often can be reversed for convenience in connecting to operating mechanism. Dynamic spool seals can be externally replaceable (Figure 13.14). The individual load checks in the supply line prevent the load from dropping when changing spool positions and prevents back filling from one cylinder port to another when operating two spools simultaneously. Spring-to-neutral positioning can be replaced with detent positioning.

Many options for actuation of the valve spool are readily available (Figure 13.15). Lever actuation may be required for certain positioning functions, while remote linkage may be desirable for other applications. A sectional view of the machined faces and O-ring static seals is shown in Figure 13.16.

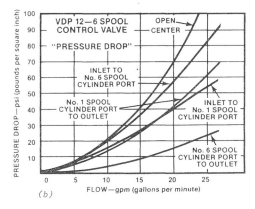

Figure 13.12 Typical six-spool valve with operating data. (a) External view of the monoblock design with integral externally adjustable relief valve. (b) Typical pressure drop values available from the manufacturer to estimate pressure drops when trouble shooting a machine. Courtesy of Parker Hannifin Corporation Mobile Hydraulic Division, Cleveland, Ohio.

External ports are machined to accept SAE (Society of Automotive Engineers) standard straight thread fittings that provide leak-free connections with provisions to accurately position elbow-type fittings. A flow control function is also provided in the assembly of Figure 13.17. Optional manual and electrohydraulic pilots are shown in Figure 13.18. Four-bolt flange connections simplify interconnections in nominal pipe sizes of 1 in. or larger (Figure 13.19).

REMOTE PILOT CONTROL

Lever operation of valve spools provides an excellent feel for the machine operator. At times the connection to the valve spool can be difficult with mechanical devices and a pilot or electrical remote control may provide easier access and adequate sensitivity of control. The joy stick control of Figure 13.20a consists of four offset pilot valves. The pilot assembly can apply pressure at a value established by the operator to the cavity at either end

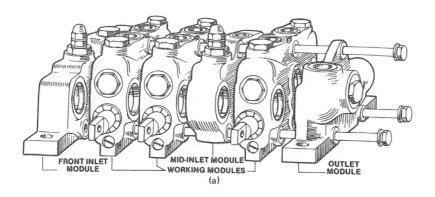

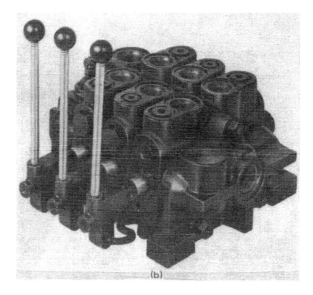

Figure 13.13 Segmented control valve assembly. (a) Mid-inlet module permits flexible incorporation of additional pump capacity. (b) Lever-operated model with anti cavitation checks. Courtesy of Parker Hannifin Corporation Mobile Hydraulic Division, Cleveland, Ohio.

of the pilot operated, spring-centered, spool valve. The pilot pressure is balanced by the centering spring within the piloted valve. As pressure is decreased by action of the remote pilot valve, the piloted spool is pushed to center. As pressure is increased the piloted spool moves proportionally to direct the desired flow to the actuator whether it be a linear cylinder or hydraulic motor. The operator soon becomes accustomed to the control and the results are quite similar to the direct lever operation. With four modules in the joy stick it is possible to operate two directional valves for lift and swing, etc. Figure 13.20b illustrates typical pilot units. A similar joy stick structure can be fitted with four electrical controls to actuate a proportional solenoid or a pulse width modulation unit (Figure 13.21). Figure 13.22 illustrates a directional control unit designed specifically for a farm tractor.

Figure 13.14 Dynamic spool seals can be replaced from the exterior without major disassembly of the valve segments. Courtesy of Commercial Intertech.

Figure 13.15 Typical 16 gpm at 3000 psi rated mobile valve with operational options appropriate to machine needs. Cartridge format eliminates piping needs for pressure control functions. Courtesy of Husco International, Inc.

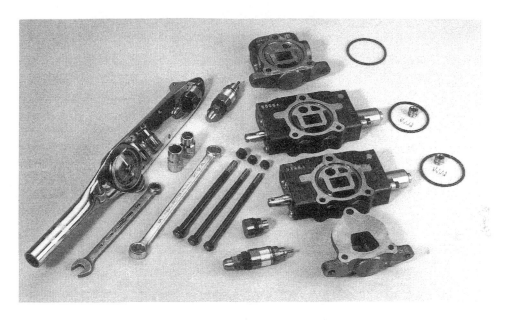

Figure 13.16   Valve sections disassembled showing O-ring-type gasket. Courtesy of Husco International, Inc.

Figure 13.17   Sectional valve with rating of 35 gpm at 3600 psi. Integral flow control function is included. Courtesy of Husco International, Inc.

Figure 13.18   Multiple-section valve with optional manual and electrohydraulic pilots. Courtesy of Husco International, Inc.

Figure 13.19   Four-bolt flanges simplify connections to the mobile-type control valve assembly. Courtesy of Husco International, Inc.

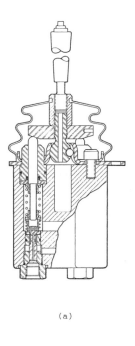

(a)

(b)

Figure 13.20 (a) Joy stick has lever and four metering valves to provide remote control of two hydraulically-piloted directional control valves. Positive pressure feedback helps operator sense speed of operation. Hollow handle permits wiring to thumb switch at top. (b) Optional remote control assemblies. Courtesy of Husco International, Inc.

INTEGRAL PILOT LINE FILTER

Precision pilot and feedback devices require clean power conducting fluid. Because of the possibility of inadvertent contamination of fluid in major lines a pilot filter may be incorporated in the pilot line to the pilot stage. Note the pilot filter in the sectional view of Figure 13.23. Pilot line filters are usually of relatively large capacity so that they are not normally serviced unless a catastrophic contamination of the major fluid lines have occurred. They must be serviced in a clean atmosphere to avoid contamination downstream of the filter with associated potential for malfunction of the precision pilot.

PROPORTIONAL SOLENOID CONTROL VALVES

The proportional control valve of Figure 13.24 employs a piloted assembly (7) controlled by a pilot valve (3). A piloted spool (8) is machined to provide appropriate passages to an actuator. A fluid motor or double-rod cylinder with equal displacement in either direction can employ a piloted spool as shown with equal lands at port A and port B. Machining of the spool land at port A or B can be modified to compensate for the rod displacement for a

Figure 13.21   A remote joy stick operator with appropriate electrical control facilities can be used to actuate a valve designed for pulse width modulation actuators or proportional solenoids. Courtesy of Husco International, Inc.

Figure 13.22   Valve bank designed for controlling an agricultural tractor. Courtesy of Husco International, Inc.

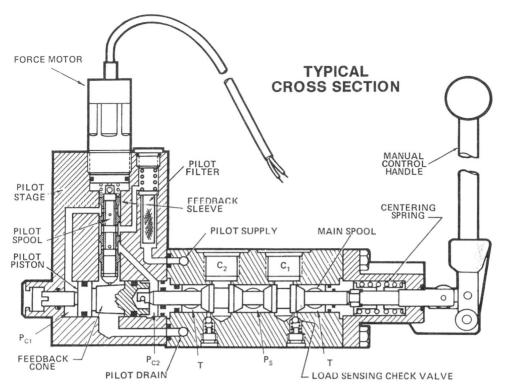

Figure 13.23 Directional control valve with force motor actuation and manual override. Courtesy of Dynex/Rivett, Inc.

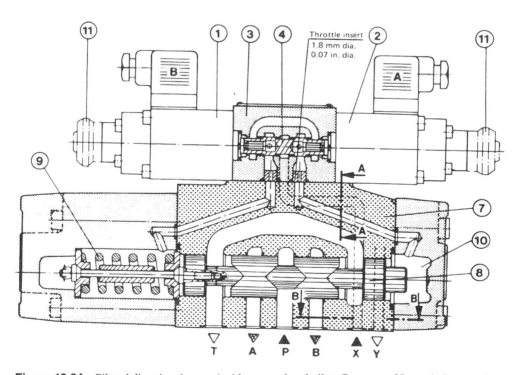

Figure 13.24 Piloted directional control with proportional pilot. Courtesy of Rexroth Corporation.

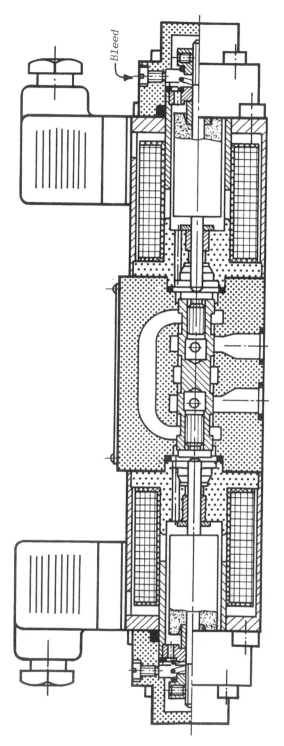

Figure 13.25 Proportional pilot valve with integral pressure feedback pilot pistons. Courtesy of Rexroth Corporation.

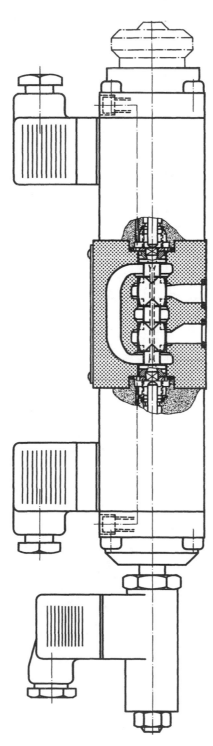

Figure 13.26 Proportional directional control with direct solenoid actuation and Linear Variable Differential Transformer (LVDT) feedback assembly. Courtesy of Rexroth Corporation.

single-rod cylinder. If the spool or bore has been damaged and a replacement is needed it must be the same as the original and installed in the same physical position to retain the original characteristics of the flow through the valve. The proportional valve may be designed to provide both flow direction and flow rate. Compensation may be included in the circuit design so that the proportional spool serves as both a flow and directional control valve in a dedicated flow pattern.

The bias spring (9) serves to center the spool (8) when the pilot valve (3) is not actuated by either solenoid (1) or solenoid (2). Manual actuation can be accomplished by pushing the mechanical actuator (11). Energy flow to solenoid A will move the spool (4) and direct pilot flow to the cavity of spring (9). Resistance created by the movement of the spool and compressing of spring (9) is reflected to the small internal piston at the solenoid B end of pilot spool (4). Figure 13.25 provides an enlarged sectional view of the piston within the spool (item 4, Figure 13.24). In operation the internal piston tends to move the spool to the neutral position, blocking the flow to the cavity at the spring (9), and the piloted spool (8) stays in the partially shifted position until additional energy is directed or released from solenoid A. The position of piloted spool (8) is proportional to the energy flow to solenoids A or B within the design parameters of the valve assembly. Note the bleed screw at each end of the solenoid of Figure 13.25. At initial installation and/or after a repair has been

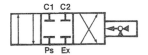

Figure 13.27 Servo valves provide precise control and repetitive accuracy. Available for pressures to 5000 psi (345 bar). Courtesy of Schenk Pegasus Corporation.

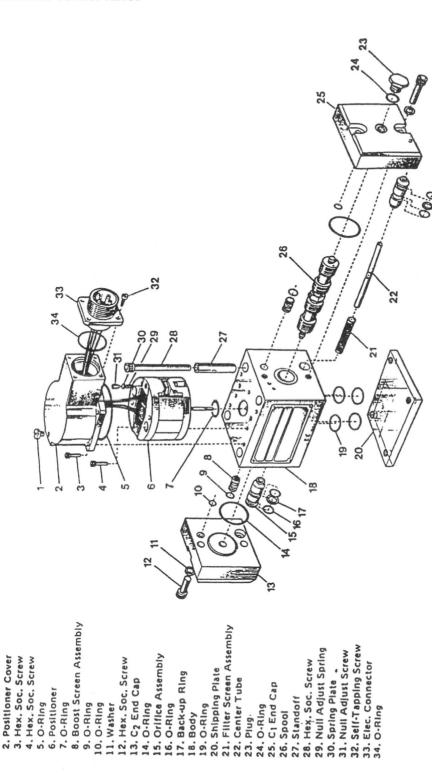

1. Seelskrew
2. Positioner Cover
3. Hex. Soc. Screw
4. Hex. Soc. Screw
5. O-Ring
6. Positioner
7. O-Ring
8. Boost Screen Assembly
9. O-Ring
10. O-Ring
11. Washer
12. Hex. Soc. Screw
13. C_2 End Cap
14. O-Ring
15. Orifice Assembly
16. O-Ring
17. Back-up Ring
18. Body
19. O-Ring
20. Shipping Plate
21. Filter Screen Assembly
22. Center Tube
23. Plug
24. O-Ring
25. C_1 End Cap
26. Spool
27. Standoff
28. Hex. Soc. Screw
29. Null Adjust Spring
30. Spring Plate -
31. Null Adjust Screw
32. Self-Tapping Screw
33. Elec. Connector
34. O-Ring

Figure 13.28 Typical exploded view of a precision hydraulic servo valve. Courtesy of Schenk Pegasus Corporation.

completed the entrapped air must be exhausted from each solenoid prior to resuming normal actuation.

LVDT FEEDBACK

The linear variable differential transformer (LVDT) attached to the flow directing spool assembly of Figure 13.26 monitors the position of the spool to ensure repetitive accuracy. Spool design may be modified to provide the desired directional control and flow rate. Valve or spool replacement must equal the original design.

SERVO VALVES

Servo valves are manufactured in many different configurations with feedback options. The high amplification capabilities require precision machining and assembly, and the fluid medium must be maintained at a surgically clean level. Most manufacturers recommend repairs be assigned to factory trained personnel with clean room facilities. Note the symbol in Figure 13.27 indicating a closed center condition. The rectangle at the right end of the symbol indicates that the actuator is a force motor with feedback control capabilities. Figure 13.28 shows an exploded view of the servo valve of Figure 13.27.

14
Check Valves

BASIC CHECK VALVES

The swing check valve of Figure 14.1 is primarily used in processing systems with water, gas, air, etc. Pressures may be limited to a few hundred pounds per square inch. Check valves for higher pressure service may be as simple as a ball and seat structure. A bias spring is often included to permit installation of the check valve in any plane. Note the bias spring in the angle check valve of Figure 14.2. Flange piping connections may be used in larger size valves as shown in Figure 14.3. A symbol for a check valve is shown below Figure 14.3a. The symbol below Figure 14.3b indicates a spring if it is desirable to indicate on a drawing that the check valve includes a bias spring. The check poppet of Figure 14.3b includes a removable pipe plug that can be drilled to a desired diameter to provide a controlled flow in one direction and a free flow in the other direction. In line check valves can be designed with female ports (Figure 14.4a) or male ports (Figure 14.4b).

THE ELASTOMERIC SEAL

The check valve of Figure 14.5 is designed to fit into a hydraulic tubing line without need for flare or threads on the tubing line in which the valve is installed. The elastomeric seal provides an excellent closure at either high or low pressures. The check of Figure 14.6 is designed for pressures of approximately 200 psi or less. The sealing function is provided with an O-ring assembled in a fitted groove in the nose piece to permit flow around the O-ring in one direction with the reverse flow sealed by the ring. The needle-type flow control of Figure 14.7 can be located concentrically in the check valve body to provide a free flow in one direction and an externally adjustable flow in the other.

THE SLIP IN CHECK VALVE ASSEMBLY

A captive elastomeric seal with a back-up ring provides the flow barrier for the outer diameter of insert-type check valves (Figure 14.8). The insert-type format eliminates the

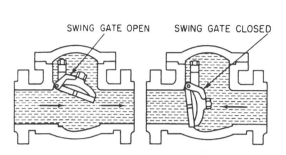

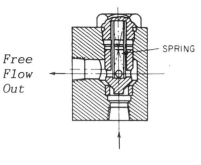

Figure 14.1 Swing check offers little resistance to flow for low pressure service.

Figure 14.2 Angle check valve with bias spring and threaded ports.

need for external connections. A smooth bore compatible with the resilient sealing assembly is all that is needed to accommodate the slip-in-type cartridge check. The check valve assembly of Figure 14.9 also incorporates a captive elastomeric ring at the seal point in the check valve. A collar-type seal is employed in the insert-type check valve of Figure 14.10. Insert-type check valves can be incorporated in host devices without need for conventional piping. Intake and exhaust insert-type checks are illustrated in the reciprocal pump mechanism of Figure 14.11a. The check valve of Figure 14.11b is employed to prevent back flow through the gear type pump. The insert-type check valve of Figure 14.11c prevents back

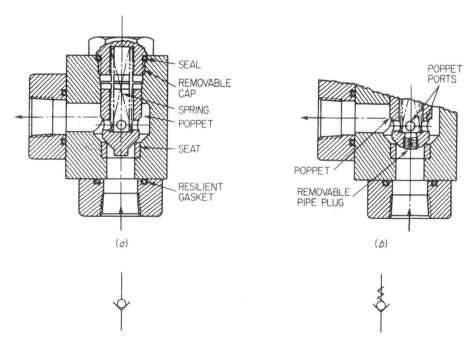

Figure 14.3 Angle check valve with bias spring and flange connections. (a) Full closure for return flow. (b) Removable plug which can be drilled to desired size for fixed orifice to provide free flow in one direction and a restricted flow in the opposite direction.

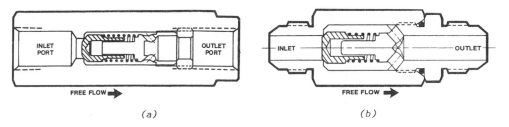

Figure 14.4 In-line check valves: (a) with female ports, (b) with male ports. Courtesy of Fluid Controls, Inc.

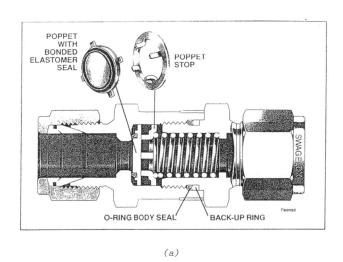

(a)

Operation

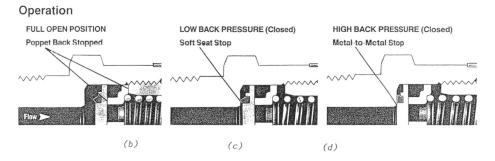

Figure 14.5 In-line check valve with elastomer seal: (a) sectional view, (b) open position, (c) low pressure seal, (d) high pressure seal with metal-to-metal stop. Courtesy of Crawford Fitting Company.

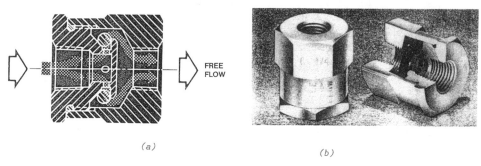

(a)   *(b)*

Figure 14.6 Low pressure check valve with elastomeric seal: (a) flow path, (b) external and sectional view. Courtesy of Ace Controls, Inc.

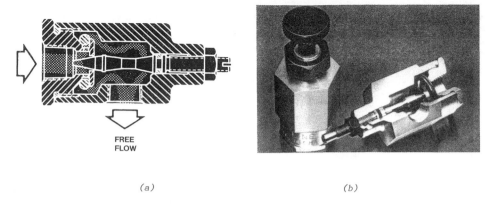

Figure 14.7 Flow control and check valve. (a) The needle valve assembly is located concentrically in the check valve body. (b) External and sectional view. Courtesy of Ace Controls, Inc.

flow through the directional control valve. The insert-type check valve of Figure 14.11d prevents back flow through the manifold assembly. The plastic check valve of Figure 14.12 can be installed in either position within the proposed cavity for desired free flow direction. Electrical conductivity is minimized with the plastic assembly.

An insert-type cartridge in a block designed to sandwich between conventional flanges minimizes piping (Figure 14.13a). A threaded insert-type format also minimizes piping and installation space (Figure 14.13b). Typical check valve inserts are illustrated in Figure 14.13c. The insert-type sandwich flange is shown in Figure 14.13d.

PILOT OPERATED CHECK VALVES

Pilot operated check valves are an extension of direct operated check valves; these valves simply allow flow in one direction and stop flow in the opposite direction. The pilot operated check valve provides control versatility to the circuit designer by being able to control flow in either direction on command. Pilot operated check valve must completely shut off flow with zero leakage on command. In the opposite direction they must allow virtually unrestricted flow within the design parameters of the system.

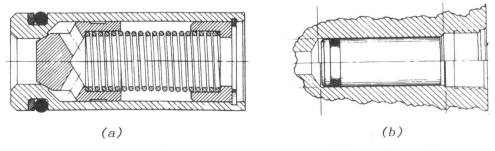

(a)   (b)

Figure 14.8 Insert-type check valve with metal-to-metal seat. (a) Sectional view of poppet and spring assembly, (b) sectional view of valve in bore.

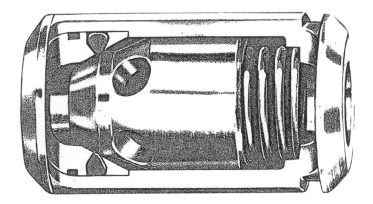

Figure 14.9   Insert-type check valve with resilient elastomeric seat. Courtesy of Kepner Products Company.

System Requires Valve to be Closed

When the pilot operated check valve is used to "hold" the load and actuator creep occurs one of several problems exist:

> The pilot system is misapplied or poorly designed and pressure causes the valve to open.
> Internal leakage results from a worn valve seat or main poppet.

The second condition is generally the problem; however, checking pilot pressure at gauge TP3 (Figure 14.14) first could be a time saver.

Internal leakage most generally is the result of wear. Additionally, scoring of the main poppet or poppet bore due to fluid contamination can cause the main poppet from seating correctly.

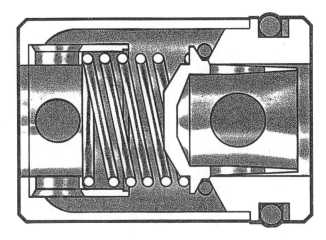

Figure 14.10   Insert-type check valve with O-ring-type collar seal. Courtesy of Circle Seal Controls, Anaheim, CA.

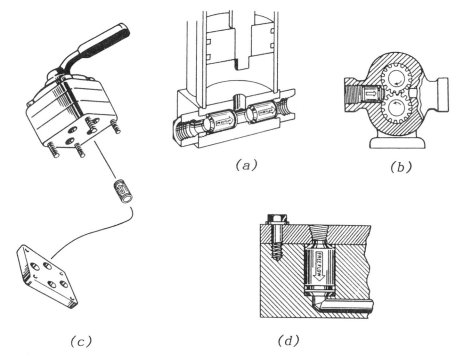

Figure 14.11 Insert-type check valves. (a) Plunger pump concept. (b) Back flow prevented through gear pump. (c) Back flow prevented through directional valve. (d) Back flow prevented in manifold assembly. Courtesy of Kepner Products Company.

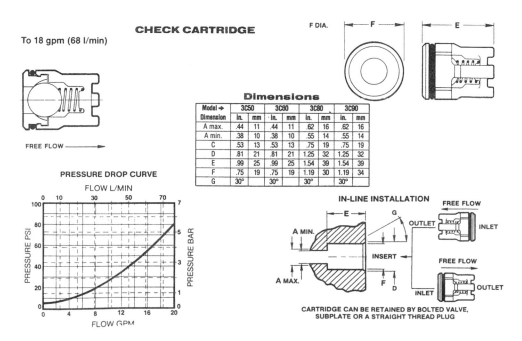

Figure 14.12 Plastic insert-type check valve with typical installation information. Electrical isolation is possible. Courtesy of Fluid Controls, Inc.

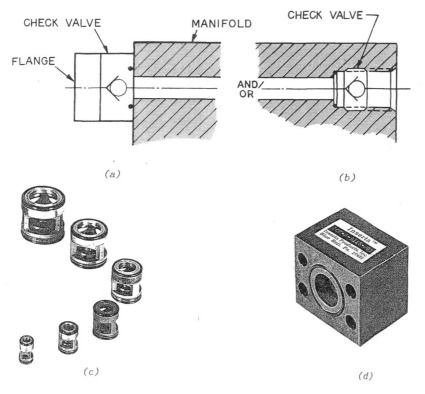

Figure 14.13 Insert-type check valves in flange and threaded format. (a) Flange insert. (b) Threaded insert. (c) Typical threaded insert-type assemblies. Courtesy of Inserta Products, Inc.

System Requires Valve to be Open

When the pilot operated check valve is required to be open to allow free flow through the valve while the fluid upstream of the valve is under load, pressure at gauges TP1 and TP2 should read the same. If this is not the case, several conditions could be causing the problem:

There is insufficient pilot pressure to actuate the valve pilot spool.
The main poppet is frozen shut and will not move.
The pilot piston will not move.

Both the second and third conditions can be due to contamination that hinders movement by scarring of the main poppet/pilot piston or their bores.

When using a pilot-to-open check valve the ratios of the pilot piston area to the cylinder areas associated with the pilot operated check valve must be correctly proportioned. The ratio between the pilot piston area and the main poppet in the valve must be greater than the ratio between the cylinder piston area and its annulus area in the rod end. A good condition would be for the cylinder piston annulus area to be 2:1 with the check valve area. The establishment of the proper check valve area ratio is the responsibility of the valve designer. The maintenance person needs to understand this area relationship when called upon to troubleshoot the circuit. Once the problem is isolated to be the pilot operated check valve the repair or replacement action must be taken.

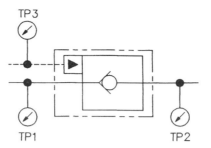

Insert Gages at TP1, TP2 and TP3

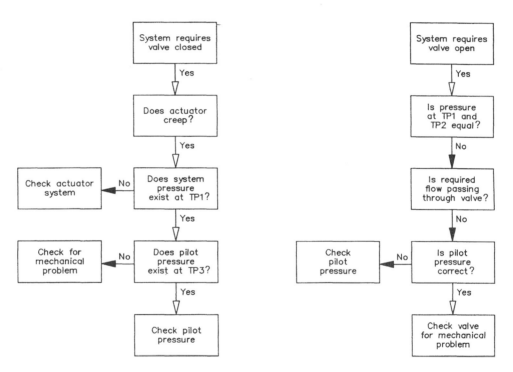

Figure 14.14 System test for pilot operated check valves. Courtesy of Vickers, Incorporated.

PILOT OPERATED CHECK VALVES

Pilot-to-Close Check Valves

Pressure at port A of the check valve in Figure 14.15a can pass to port B at the pressure value created by the bias spring and the effective area of the poppet face. Pressure at port B can pass to port A after overcoming the installed value of the bias spring as pressure pushes the poppet open with the effective shoulder area of the poppet. Pressure at the pilot adds to the value of the bias spring so that the valve will stay closed until pressure at port A or port B is great enough to force the poppet from the seat. Figure 14.15b illustrates a threaded cartridge valve assembly that incorporates a spring to hold the poppet from the seat when pilot pressure is released. Models are available with a bias spring to hold the screw-in-type cartridge valve poppet normally closed (Figure 14.15a).

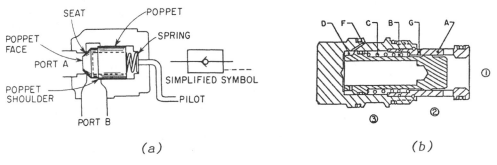

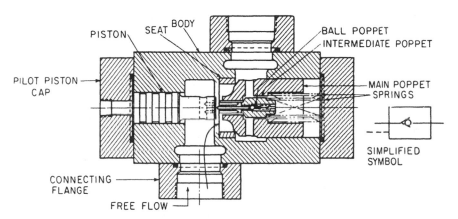

Figure 14.15 Pilot-to-close check valve. (a) Pressure at the pilot will add to bias spring force to maintain the poppet in the closed position. Simplified symbol shown at right. (b) Pilot pressure at port 3 is directed through passage F to urge poppet against bias spring C in the closing function. Courtesy of Sun Hydraulics Corporation.

Figure 14.16 Pilot-to-open check valve with three-stage operation. Ball poppet to release highest pressure. Intermediate poppet for secondary pressure release. Main poppet provides major flow path. Simplified symbol at right.

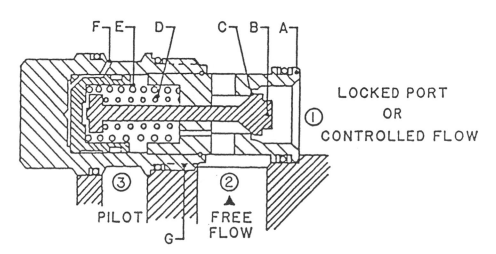

Figure 14.17 Screw-in cartridge-type pilot operated check valve. Pilot pressure is directed through passage F to urge poppet C from the seated position. Free flow is from port 2 to port 1. Courtesy of Sun Hydraulics Corporation.

Pilot Operated Check Valves with Multiple Stage Pressure Release

A pilot-to-open pilot check may be used to lock high pressures in an actuator during a hold portion of a machine cycle. To avoid shock conditions as the high pressure is released, the pilot piston of Figure 14.16 unseats a small ball poppet. As the pressure decays the intermediate poppet opens and a greater flow is provided through the valve. At a relatively low pressure the major poppet opens for a free return of the remaining captive fluid through the pilot check valve. A simplified symbol is shown adjacent to the cross section view.

Cartridge-Type Pilot Operated Check Valve

The cartridge-type screw-in pilot operated check valve of Figure 14.17 permits free flow from port 2 to port 1 past poppet C. Bias spring D creates a modest resistance to flow. Locked pressurized fluid at port 1 can be released with an application of proper level pilot pressure at the pilot port (3). Spring E retracts the actuator piston as pilot pressure is released at port 3.

Sandwich-Type Dual Pilot Operated Check Valves

Both actuator lines may need to be locked to hold a machine member in a desired position. Two check valve assemblies with a common actuator can be assembled into a common housing (Figure 14.18) or in a similar sandwich-type assembly to fit between a directional valve, the associated subplate, and the actuator (Figure 14.19). Note the pilot source from the opposite actuator line employed to open the locked valve when the directional control valve is shifted in the machine cycle. Note the flow path in the neutral position of the four-way directional valve of Figure 14.19. The two cylinder lines are connected to each other and to the tank. A closed center four-way directional control valve may be subject to leakage in the neutral position; this leakage can cause the pilot check to open and create a creep condition in a potentially hazardous installation. This condition is most often found in new installations or where a four-way directional valve has been replaced with a closed center spool in place of the proper spool providing either an open center condition in neutral

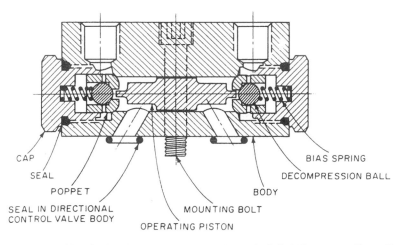

Figure 14.18 Two pilot check valves in a single housing to lock fluid in actuator lines. Courtesy of Rexroth Corporation.

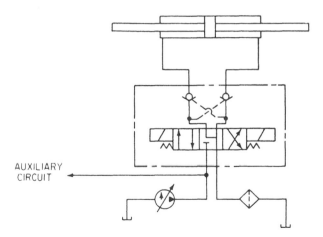

Figure 14.19 Typical application of dual pilot operated check valves. Courtesy of Rexroth Corporation.

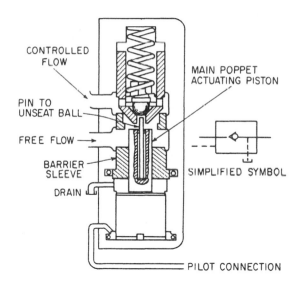

Figure 14.20 Pilot check valve with barrier sleeve to minimize flow forces during high velocity flow conditions. Simplified symbol at right.

or with the cylinder ports connected together to the tank port; the latter ensures that undesired pilot pressure is not available to the pilot port of the pilot operated check valve assembly (Figure 14.19).

Pilot Operated Check with Barrier Sleeve

The pilot check of Figure 14.20 is assembled with a barrier sleeve to prevent high velocity flow impinging on the face of the pilot piston opposite to that where pilot pressure is applied. The barrier cavity is drained so that counterforces cannot affect the piloting function except at the small diameter passing through the barrier sleeve.

15

Hydraulic Cylinders (Linear Actuators)

Hydraulic cylinders often function in a hostile environment and the potential for damage is great. The test procedures of Figure 15.1 one can uncover basic problems. Erratic movement or failure to move may be caused by piston bypass leakage or piston/cylinder bore scoring. Erratic cushion action can be caused by improperly adjusted cushions, scoring of cushion bore/cushion pilot, or faulty cushion speed control mechanism.

PROPER STORAGE PAYS OFF WITH HYDRAULIC CYLINDERS

It pays to keep spare hydraulic cylinders on hand for use when you need them, particularly for hard worked construction machinery, but repair and maintenance personnel should know and follow recommended storage practices or the cylinders can be a source of continuing problems.

Hydraulic cylinders, though often large and unwieldy, are precision machines with finely finished parts and close tolerances. And they are expensive, so handle them with care.

Effect of Environment

For optimum storage life, hydraulic cylinders should be kept in an environment that is protected from excessive moisture and temperature extremes. A hot, dry desert climate with cold nights, for example, may need to be accommodated when choosing the storage area. Daytime heat quickly bakes oil out of sealing materials which causes leaks and rapid wear when the cylinder is placed in service. Cooling at night causes water condensation and corrosion damage to wear surfaces. Storage areas that allow exposure to rain, snow, and extreme cold must likewise be avoided.

Storage

It's best to store cylinders indoors if possible. But indoors or out, be certain that plugs or closures are properly installed in all ports to keep out moisture and dirt. Over tightening of closures should be avoided. Widely varying temperatures and tightly closed ports may

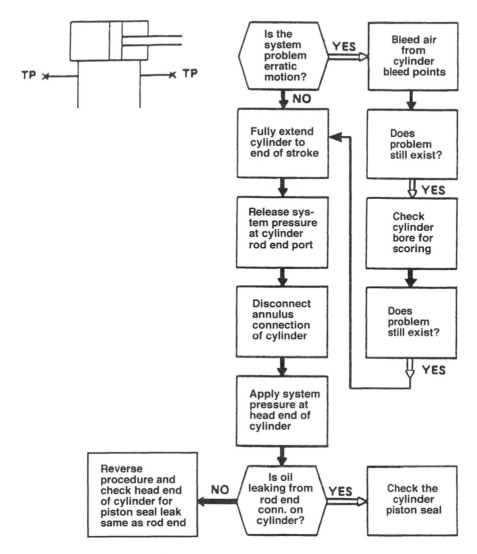

Figure 15.1 System test for cylinders. Courtesy of Vickers, Incorporated.

cause pressure to build up inside the cylinder to the point where the piston moves far enough to expose the rod to contamination or corrosion.

Choose a storage location where the cylinders are protected from physical damage. Even a little ding from a falling bar or forklift tine can cause trouble later.

Position

Cylinders, particularly large ones, should be stored closed in a vertical position with the rod end down. Be certain they are blocked securely to keep them from toppling. Storing with rod ends down keeps oil on the seals and protects them from drying out. This is more critical with rag and butyl seals than with urethane sealing materials.

Storing single-acting cylinders with the rod end up can cause port closures to pop open and leak, exposing the sleeves to corrosion damage and contamination. Storing with the rod end down also discourages the temptation to lift a cylinder by the rod eye—a dangerous practice.

If horizontal storage cannot be avoided, the rod or cylinder should be rolled into a new position every two months or so to prevent drying, distortion, and deterioration of the seals. Do not forget that a cylinder can be a major source of contamination. A small scratch or nick on the sleeve will quickly shred the packing and contaminate the system. Store cylinders carefully and keep them clean (Figure 15.2).

HYDRAULIC CYLINDER MAINTENANCE

To get maximum service life from your equipment's hydraulic cylinders, set up a planned program of periodic maintenance service (Figure 15.3). Inspect cylinders in their extended positions for nicks, weeping at packing gland and alignment with fixed and moving machine members. Early repairs can minimize potential for failure and/or major repairs.

Look for the Obvious

Items on your cylinder check list will vary with the equipment and its working environment, but any list developed should include these obvious checkpoints:

1. *Cleanliness.*   Frequently clean the entire machine to eliminate dirt buildup on cylinder rods and mountings.
2. *Condition of oil.*   Keep the oil clean to avoid accelerating normal wear. Have a sample of the hydraulic oil tested by a qualified laboratory for contamination

Figure 15.2 Changing position of cylinders stored in a horizontal position on a programmed interval can insure lubrication and avoid distortion. Courtesy of Commercial Intertech.

Figure 15.3 Periodic inspection of hydraulic cylinders can minimize potential down time. Courtesy of Commercial Intertech.

and/or deterioration. If tests indicate deterioration of the oil change the systems oil and filters. Check the filler cap and breathers regularly.

3. *Machine efficiency.* Sluggish or spongy cylinder operation, particularly in single acting cylinders, may indicate air in the system. Correct by bleeding the system.

4. *Cylinder mountings.* Check for excessive wear that may indicate lack of lubrication. Poor lubrication also increases friction on mountings and puts bending stress on the cylinder. Lubricate regularly and repair or replace worn parts.

5. *Seals.* Leakage indicates worn or damaged seals and possibly a scored or nicked cylinder rod. Replace seals and overhaul cylinder if necessary. Make certain that the replacement oil is compatible with the seals.

6. *Relief valve settings.* Follow the manufacturers recommended pressure values. Too high a setting leads to premature failure of hydraulic components and possible structural failure as well. Settings that are too low cause reduced efficiency, heat buildup, and shortened oil life. Check for proper relief valve setting.

Find the Real Problem

Don't just replace or overhaul a cylinder that has failed. Look for the real cause of failure in the cylinder itself or elsewhere in the system. The cylinder is designed to withstand full relief valve pressure. Improperly set or nonfunctioning relief valves can cause excessive stresses on seals and mountings, overheating, and early cylinder failure. Save your cylinders by periodically checking for proper relief valve settings and overall system performance. Monitor in service performance to make certain that the operator is not inducing shock loads.

Work Area

Normally, cylinder overhaul work should be done in the maintenance shop where parts can be kept clean and proper tools are available. In an emergency, it may be necessary to repack a cylinder on the machine under field conditions. In either case, take every precaution to avoid contamination from the work area. Change or clean the filter elements to eliminate contamination introduced by the damaged cylinder. Recheck the entire system next time the equipment comes into the maintenance shop.

WORKING WITH HYDRAULIC CYLINDERS

By following a few simple rules for handling and working around hydraulic cylinders you can greatly extend their service life and minimize equipment downtime. All it will cost you is a little extra care in protecting these strong, yet vulnerable, components from every day hazards in the field, factory, storage area, or in the maintenance shop.

Handle with Care

Remember, when replacing a hydraulic cylinder, careful handling can have considerable effect on its future performance. It is always a good practice to lift cylinders with a nylon

Figure 15.4 Always handle hydraulic cylinders with care. A nylon sling is not nearly as likely to scar precision surfaces. Courtesy of Commercial Intertech.

Figure 15.5 A small splatter of molten metal can ruin a cylinder rod. Use care when welding near a cylinder. Use protective covers if potential damage can be anticipated. Courtesy of Commercial Intertech.

sling instead of a chain that can easily scar polished surfaces and subsequently lead to seal failure (Figure 15.4). Cylinder barrels and sleeves can also be dented with hoist chains or falling objects, thereby impairing performance and causing components to wear unevenly.

It is always a good practice to handle cylinders in the collapsed or closed position. This helps to avoid damage to polished sealing surfaces of the cylinder rod. Remember that pin-eyes are for mounting, not lifting. If the cylinder must be lifted by a pin-eye, use extreme care to prevent damage caused by the cylinder rod pulling out of the body. A pull-out can also suck contamination into the cylinder ports to cause more trouble later. Make certain all cylinder ports are plugged during storage.

Working Near Cylinders

Extreme care must be exercised when making repairs on machine parts or systems that are near hydraulic cylinders. Never walk or stand on hydraulic cylinders and especially not on the extended cylinder rods. Be particularly careful to cover hydraulic cylinders when welding nearby (Figure 15.5). Use a cover that resists sparks and spatters of considerable size. It takes only one small spatter of molten metal on a polished cylinder rod to tear up the seals and ruin the cylinder.

Never forget the importance of keeping hydraulic cylinder rods clean and well protected from dirt, grit, abrasion, or other hazards that can impair performance or cause premature cylinder failure. Nicks or scratches on cylinder rods will cause seal failure and eventually mean replacement of the cylinder. A wiper ring can protect major seal assemblies as well as simple O-ring-type seals (Figure 15.6a). The wiper ring may be an integral

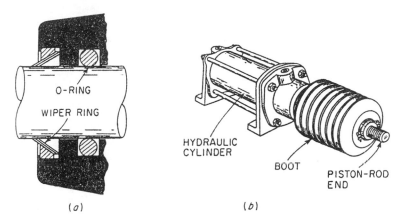

Figure 15.6 Protecting the rod packing. (a) A wiper ring can protect packing from contaminants. (b) A protective boot may be desirable in a hostile working environment.

part of the seal assembly. A protective boot (Figure 15.6b) may be necessary in an especially hostile working environment.

Assembly of Tie Rod Type Cylinders

Surfaces adjacent to the cylinder tube must be maintained in a parallel position, especially in sizes of approximately 4-in. diameter and larger (Figure 15.7). Tie rods must be tightened uniformly or the heads will not be parallel. Problems that may be difficult to analyze occur as the piston comes to rest against heads at the end of stroke. Also, seal and wear problems shorten cylinder life.

One quick way of checking is to tap the tie rods; they should "ring" at about the same frequency as judged by ear. Also, tie rods should not be twisted. Check by making a chalk line on tie rods before torquing. The line should not be spiral when done! If a rod is not straight, back-off slightly until the line is straight. If you do not have straight tie rods, the cylinder heads may wind up "twisted" after the cylinder is pressurized a few times (Figure 15.8).

CYLINDER MOUNTING

Types

Assuming that the basic cylinder design and pressure rating has been established, the next step in the troubleshooting task is to determine that the cylinder mounting type is best suited to the machine application. This involves consideration of the various mounting types that may be available. Cylinder types may be classified generally by whether or not thrust is absorbed on the cylinder centerline. Mountings where this action occurs are available in fixed mount styles and pivot mount styles (Figures 15.9a and b).

In mounting types where thrust is not absorbed on the cylinder centerline (Figure 15.5c), the thrust from the piston rod is taken up at a mounting surface that is usually parallel with, but not coincident with, the cylinder centerline.

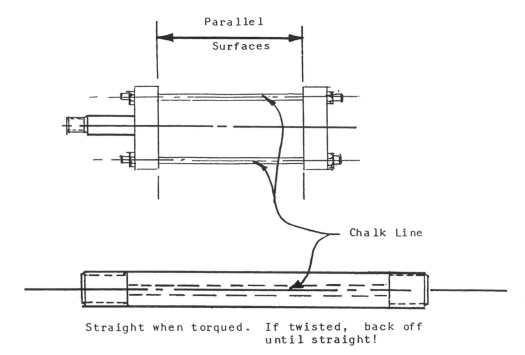

Figure 15.7 Assembly of tie-rod cylinders.

Mounting Specifications

Specification of the type of mounting for a cylinder depends largely upon the application. Some of the factors that should be considered include:

1. Cylinders with noncenterline-type mountings tend to sway under load (Figure 15.10a). The resulting additional wear on the cylinder can shorten the useful life of the component. This is especially important on larger bore cylinders.
2. The rigidity of the machine frame should be taken into account. It should be remembered that cylinders with noncenterline-type mountings often require stronger machine members to resist bending moments (Figure 15.6b).
3. If the motion of the machine member acted on by the piston rod is essentially linear, then in most cases a fixed mounting-type cylinder should be used. If the machine member moved by the piston rod travels in a curvilinear path, a pivot-

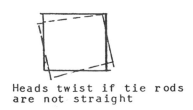

Figure 15.8 In the absence of straight tie-rods, the cylinder head may end up "twisted" after cylinder is pressurized several times.

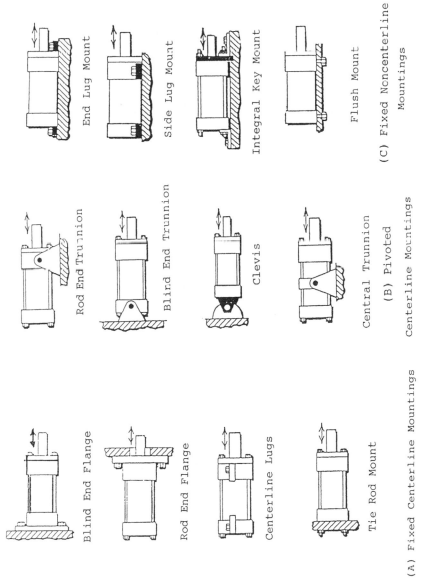

Figure 15.9 Power cylinders are available with a variety of mounting arrangements. Fixed centerline mountings (a), and pivoted centerline mountings (b), provide for piston rod thrust to be taken up along cylinder centerlines. This minimizes swaying of the cylinder and flexing of the machine members to which the cylinder is mounted. Cylinders with fixed noncenterline mountings (c) are convenient in many applications where thrust may be taken up at offset mounting surfaces.

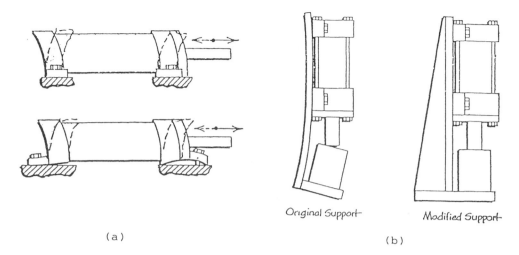

Original Support   Modified Support

(a)   (b)

Figure 15.10   Noncenterline-type mountings. (a) In applications involving large forces, cylinders with noncenterline-type mountings tend to sway under load. (b) Use of noncenterline-type cylinder mounting may require strengthening machine members to resist bending under load. In large cylinders over 6-in. diameter, swaying can create premature failure of seals, both static and dynamic.

mounted cylinder is the obvious choice. Pivot mountings are available with the pivot points on the rod end, on the blind end, or centrally located near the midpoint of the cylinder body. In most cases, a review of the rod end path will determine if a change to a pivot mounting will be more appropriate to the power transfer function.

4. The matter of cylinder strength versus stroke length should be considered. Generally, long-stroke, pivot-mounted cylinders of the center trunnion-type can use smaller rod diameters without danger of buckling than can similar cylinders with the pivot at the blind end. This can be verified by reference to stroke or buckling charts published by many cylinder manufacturers. Where long cylinders employ fixed mountings, additional supports are often required to prevent excessive sag or vibration (Figure 15.11). Fixed noncenterline-mounted cylinders with very short stroke lengths introduce another type of cylinder problem (Figure 15.12). In this case, mounting bolts are subjected to increased tension that in combination with shear forces, can over-stress the bolts.

Tension or Compression Forces

The selection of a mounting style may not have considered whether the major force applied to the machine resulted in tension or compression of the cylinder rod. The blind end flange-type mounting is best for thrust loads (rod in compression), and a rod end flange mounting is best where the rod is stressed in tension.

Alignment

Alignment problems must always be considered. If misalignment occurs between a cylinder and the machine member it moves, it may be necessary or desirable to compensate for

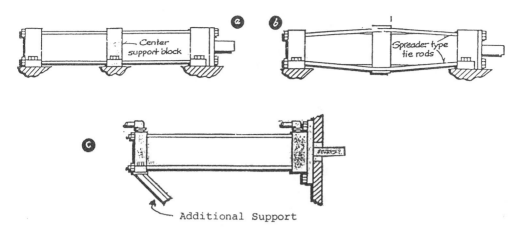

Figure 15.11 Relatively long cylinders with fixed mountings often require additional supports to prevent excessive sag or vibration. (a) Arrangement uses an extra mounting block located midway along cylinder body. (b) Spreader-type tie rods are used to increase rigidity in center portion of cylinder. (c) Where one end of a cylinder must be overhung, an additional supporting member can be provided. Care must be used in tightening tie rods as this puts column loading on the cylinder. This is worst at 0 psi. Column loads may buckle the cylinder.

this by changing the cylinder mounting. For example, a simple pivoted centerline mounting will compensate for misalignment if the misalignment is primarily in one plane. Typical mountings of this type include clevis and trunnion arrangements. Where misalignment can occur in multiple planes, then the cylinder should be equipped with self-aligning ball joints both at the blind end of a clevis mounted cylinder and in the rod end member.

INSTALLATION

The troubleshooter assumes that proper installation of a cylinder was controlled by the engineer when a machine layout was made. Drawing notes clarifying methods of installation, precautions to be observed, and so on, are normally provided for the installation personnel. Some points that manufacturing engineer should consider in the layout stage include:

1. If high shock loads are expected, the cylinder should be mounted so that full advantage is taken of the cylinder's elasticity. Using the inherent elasticity of a cylinder in shock situations can mean the difference between a successful application and one that is always giving trouble. High shock conditions usually require more rugged fastening provisions than are normally employed.

2. Fixed mounting cylinders should be held in place by keying or pinning, and some provisions should be made for this. Cylinders with integral key mounts may be used where keyways can be cut in a machine member. This type of mounting takes up shear loads, provides accurate alignment of the cylinder, and simplifies installation and servicing.

3. The use of separate keys to take shear loads is relatively common. Shear keys should be placed at the proper end of a cylinder—at the rod end if major shock load is in thrust, or at the blind end if major shock loads are in tension (Figure 15.13). In a given installation, only one end of a cylinder should be keyed to the machine. If both ends are

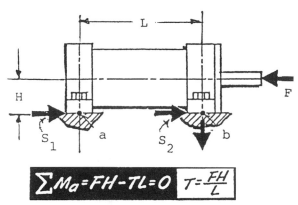

$$\sum M_a = FH - TL = 0 \qquad T = \frac{FH}{L}$$

F - Load with rod extending, pounds

H - Distance from mounting surface to cylinder centerline, inches

L - Distance between centers of rod and blind end head mounting bolts, inches

S_1, S_2 - Shear forces

T - Tension in rod end head mounting bolts, pounds

Figure 15.12 Relatively short, fixed noncenterline-mounted cylinders can subject mounting bolts to large tension forces that, in combination with shear forces, over stress the bolts. Condition shown for a cylinder rod extending against a load illustrates manner in which tension is developed in rod-end head mounting bolts. Note from the formula that, for a constant load, tension increases with decrease in distance L between rod and blind-end head mounting bolt centers. Similar analysis can be applied to determine stressing of blind-end head mounting bolts by substituting for load F the rod tension force developed during retraction, and by equating moments about point b to zero. In some cases, where large-bore cylinders have a thin-wall cylinder tube, the bolt loading and internal pressure may cause the cylinder body to enlarge in diameter near the center of its length. This can cause piston seal problems and fatigue the cylinder wall.

keyed, the advantages of cylinder elasticity in absorbing shock will be lost. Temperature and pressure effects should also be considered. A cylinder that is mounted at room temperature and zero pressure may be subjected to working temperatures of 120–140°F or more, and pressures of 2000–3000 psi or more if it is a hydraulic cylinder. Under these operating conditions it is obvious the cylinder length will increase over its installed length. The cylinder should *never* be mounted so that it is not free to expand and contract. An air cylinder (18-in. stroke length, brass body) will elongate 0.021 in. for a 100°F temperature rise. Tie rods will elongate about 0.015 in. for the same use, depending upon mount, and could be a source of problems. Even without the presence of serious shock, temperature, and pressure effects, use of shear keys at both ends of a cylinder is *not* advisable because of the difficulty of manufacturing a cylinder to an exact over-all length so that it fits between keys located for another cylinder.

4.   Locating pins may be used instead of shear keys to help absorb shear loads and to obtain cylinder alignment (Figure 15.14a). Cylinders should be pinned at either end (not both ends) for the reasons mentioned previously. Occasionally a designer will carry over

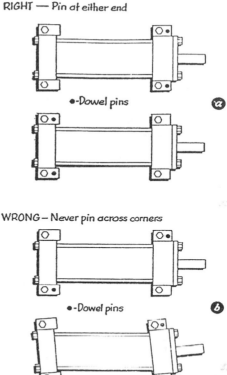

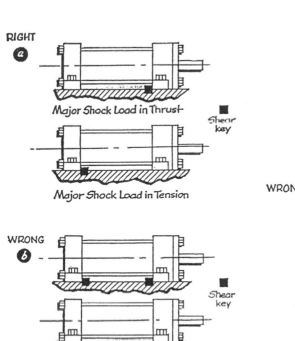

Figure 15.13 Shear keys are often used to absorb shear forces developed at cylinder mounting surfaces. Proper placement of shear keys depends upon direction of major shock loads (a). Shear keys should never be mounted at both ends of a cylinder (b). Otherwise, shock absorbing capabilities of cylinder elasticity can be lost and changes in cylinder length due to temperature and pressure effects can cause problems. Manufacturing tolerances for similar components could also make replacement difficult with such arrangements.

Figure 15.14 Cylinders that are pinned in place to help secure alignment and resist shock loads should be pinned at either end, but not at both ends (a). The choice of end depends upon direction of major shock load. If dowel pins are used across corners (b), the cylinder may be warped by operating temperatures, pressures, or shock loads.

die design experience and call for pinning across corners (Figure 15.14b). This can result in severe warping of a cylinder when it is subjected to operating temperatures and pressures.

5. Pivoted mounts should have the same type of pivots at the cylinder body and the rod end. If a simple pivot pin mount is used, the pivot pin axes at each end should be parallel. If a ball joint or self-aligning bearing is used at one end, a similar device should be used at the other. Trunnion mounts are generally designed to resist only shear loads at full-rated pressure. Therefore, self-aligning mounts should not be used to support the trunnions if the cylinder is to be used near full-rated pressure, since bending forces can also be set-up (Figure 15.15).

RIGHT   WRONG

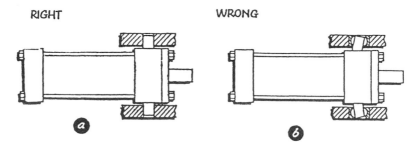

Figure 15.15 (a) Trunnions for pivot-mounting of cylinders are generally designed to resist shear loads only. (b) Use of self-aligning bearings that have small bearing areas acting at a distance from the junctions of the trunnions and the cylinder head introduce bending forces that can over stress the trunnions.

6. Two types of misalignment can occur when a fixed mount cylinder is located so that its centerline does not coincide with that of the actuated machine member (Figure 15.16). While such a cylinder may tolerate misalignment that increases with the stroke, it cannot tolerate misalignment that remains constant throughout the stroke. Section 6.3.2 of the Joint Industry Conference (JIC) hydraulic standards states "Cylinders shall be located in alignment with work slides, and shall be such that no side or radial load shall occur on piston rod or ram, unless other suitable provisions are made to take such loads." This recommendation is a good rule to follow in mounting a cylinder. In some cases, misalignment is inevitable. When this occurs, an expedient is to mount the cylinder in the manner shown in Figure 15.17. With this arrangement, oversize holes in the side lugs of a rod-end cylinder head allow the rod end of the cylinder to move up and down with respect to a pair

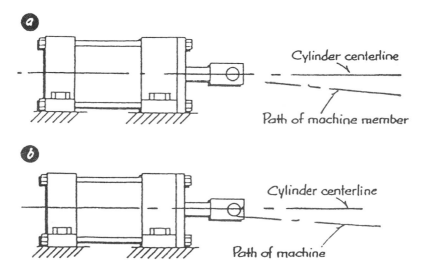

Figure 15.16 Misalignment of fixed-mounted cylinders with work slides can be of two types. (a) The cylinder can tolerate slight misalignment that increases with stroke. (b) It cannot operate properly with constant misalignment.

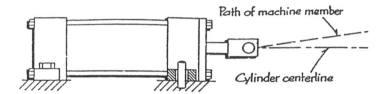

Figure 15.17 Sometimes a relatively long-stroke cylinder can be made somewhat self-aligning by allowing the rod-end head to float. In the arrangement shown, holes in the side lugs at the rod-end cylinder head permit some movement of the front of the cylinder with respect to dowel pins. The cylinder body flexes slightly about the fixed rear mounting.

of dowel pins. The blind end head is fastened rigidly to the machine, and flexure is provided by the cylinder body.

7. In attaching a cylinder rod to a machine member, the rod should not be rotated unnecessarily, as there is always danger of scoring the cylinder body. This is especially so if there is misalignment when the rod is rotated.

8. To facilitate gland maintenance rod end knuckles, cam surfaces, and other projections should not be welded or otherwise fastened permanently to piston rods during installation or after major repairs. Another point related to maintenance is that when seals need replacement the rod bearing probably should also be replaced. Seal wear rates can increase greatly when glands are oversize or worn.

9. Wherever possible, threaded joints should be pulled down tight against shoulders. This minimizes bending at necked-down sections and decrease fatigue stresses caused by reversal or forces in bending.

10. The effects of dirt and other foreign material should be considered at the installation or replacement of a cylinder. The installation and repair team should be reminded to take care that no paint or similar material is sprayed on an exposed cylinder rod, and damage to exposed rods should be avoided.

ROD-END JOINTS

In specifying a rod-end joint, the engineer should carefully consider the thread to be used at the piston rod end. For example, the thread O.D. should be somewhat smaller than the full diameter of the piston rod to allow replacement of seals without scraping their sealing edge(s) on the threads. Provision of thread relief, rather than run out of threads, provides a stronger rod end and reduces notch sensitivity. The advantages of female threads should be considered both at the design stage and in a rebuild program. Such threads are generally easier to service and if breakage occurs, a higher strength stud can be easily reinserted. In general, fine pitch threads are stronger than coarse threads because of greater total thread root area.

The type of connection of the rod end to the machine member should also be considered. The ideal rod end connection is threading the piston rod to a well-guided machine member. This arrangement gives the best possible support of the free rod end. Pivoted connections to moving machine members are usually not as critical on alignment as connections that are rigidly made to the members.

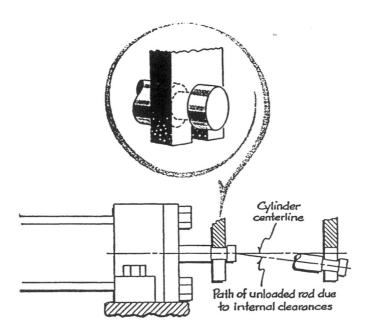

Figure 15.18 The type of rod-end connection shown—generally considered to be self-aligning—can introduce a side load on a cylinder. During extension, the piston rod end may follow the cylinder centerline because of resistance of the machine member and friction at the connection. During reversal, however, the load is relieved momentarily and the rod end can drop to the lower position. On the return stroke, side load—equal to the sum of the product of coefficient of friction at connection and the horizontal pull force—is set up.

Some rod end connections that are sometimes considered to be self-aligning cannot, in fact, provide such action. An example of this is the slotted arrangement of Figure 15.18. Here, the reduced-diameter section of a rod end engages a slot provided on a machine member that moves in a straight line. Because of necessary clearance at the connection and normal clearance between the piston and the cylinder bore, the rod end follows the lower path when no force is being transmitted to the machine member. If the machine member offers sufficient resistance, the rod follows the upper path during the outstroke of the piston rod. However, during reversal of the piston rod the force at the connection is relieved momentarily and the rod end drops to the lower position. During the remainder of the return stroke a side load is imposed on the piston rod. The magnitude of this side load is a function of the horizontal pull force required to return the load and the coefficient of friction between the rod end and the slotted member. Scoring of the cylinder body could be one result of such a side load.

PISTON ROD DIAMETER

Initial selection of the proper piston rod diameter for a fluid power cylinder is based on several factors including buckling and bearing load conditions, type of rod-end joint to be used, and circuit considerations. The troubleshooting function may indicate that some factors were not adequately considered. Typical points to remember are:

1.   The stroke length of a cylinder may be limited by rod buckling considerations. Here, the rod acts as a long column with one end semifixed in the cylinder and the other end fixed or pivoted to the machine member. Where rod buckling can occur, a feasible way to increase the stroke length is to go to a larger diameter piston rod unless the designer and manufacturing personnel are willing to accept operation at a lower design factor. Trouble at this point may indicate an increased load on the machine or a poor initial design decision. A larger rod may be mandatory to insure continuous production as a part of the repair and/or replacement activity.

2.   The stroke may be limited by allowable rod end bearing load or by bending the threads at the junction of the piston and piston rod with the rod fully extended. Here, the use of a cylinder with a stop tube may help. The stop tube increases the minimum distance between the rod end bearing and the piston.

3.   Vibration and misalignment can require the use of cylinders with large rod diameters. Usually, the bearing area at the rod-end gland is increased with larger diameter rods. Sometimes the rod-end joint determines the rod diameter selected. For example, if a large shoulder is required at the end of the rod because of high bending forces, only a large diameter rod will suffice. Special rod-end joints may also require large diameter rods.

CIRCUIT CONSIDERATIONS

Circuit considerations can affect the choice of rod diameter. The most common example of this is the use of a differential circuit. Usually, the cross-sectional area of the piston rod is about one-half the area of the cylinder bore. In some cases, the required speed or force ratios for advance and retract strokes call for other cross-sectional area relationships. This rod size relationship must be considered when replacing a damaged cylinder.

CUSHIONING

Pneumatic and hydraulic cylinders are frequently equipped with cushioning devices. One function of a cushion is to provide absorption of the energy of moving masses at the end of the stroke. This includes the masses of the piston and rod, the load which is being moved, and the fluid medium operating the cylinder. Although it might seem the energy content of the fluid medium is relatively small, it should be remembered that the kinetic energy of a moving mass is proportional to the square of its velocity. Consider, for example, a hydraulic cylinder having a 6-in. bore being fed by a line with a 1-in. bore. Whatever the speed of the piston, the fluid in the line is moving at a velocity 36 times that in the cylinder. Therefore, the kinetic energy of a given mass of fluid in the line is almost 1300 times that for the same mass of fluid in the cylinder. This kinetic energy, as well as that of the fluid in the cylinder, the piston and rod, and the load being moved, must be absorbed by the cushion in stopping the mechanisms.

From the standpoint of the cylinder manufacturer and the user, cushions can also prevent cylinder damage. This is accomplished by the removal of energy over a sufficient length of cushion stroke so that peak forces and pressures will not build-up in sufficient magnitude to damage a cylinder. An incorrectly adjusted throttle in the cushion assembly can negate the value.

A standard cushion will reduce the speed over the last portion of a stroke, but is not designed to provide reliable secondary speed control. Difficulties may be encountered, for example, if very slow primary stroke speeds are involved, since cushions are generally designed to absorb relatively high energy levels in limited time intervals. A cushion may be relatively ineffective or erratic in smoothly and consistently reducing a low primary stroke speed. Also, cushioning of pneumatic cylinders often introduces special problems characteristic of the fluid medium used.

Bounce is an effect which is caused by the absorption of the kinetic energy of moving masses without completely exhausting the compressed air that has provided cushioning action. At the moment that motion stops, some air may be stored under elevated pressure and tends to reverse the motion of the piston. This phenomenon may occur several times in rapid succession at the end of a piston stroke. The usual remedy for piston bounce is to adjust the cushion controls so that the cylinder completes its maximum stroke at the instant that all of the kinetic energy has been absorbed.

The effect of unswept volume should be considered in selecting a cylinder where cushioning is a problem. For example, a cylinder with long cushion engagement may not have the same energy absorbing capacity as a similar cylinder equipped with a short cushion. This occurs where large unswept volumes (with long cushions) reduce the cushioning action. In general, the greater the unswept volume the lower will be the compression ratio for a given cushion stroke length.

The energy absorbing capacity of a pneumatic cylinder cushion can often be increased by building up additional back pressure with a meter-out flow control valve, by decreasing the unswept volume of the cushion stroke, and by increasing the cushion stroke length if unswept volume is not increased materially.

Cushioning of hydraulic cylinders introduces problems somewhat different from those encountered with pneumatic cylinders. Bounce effects are virtually nonexistent with hydraulic cylinders, since the compressibility of a liquid medium is usually not large enough to cause trouble.

As cushioning begins for the hydraulic cylinder, initial shock may be controlled by suitable entrance tapering on both the male and female cushion parts. Also, the compressibility of a liquid medium is usually sufficient to provide some shock absorption at the beginning of cushioning action.

Machine requirements dictate the type of cushioning action to be provided. The common needle valve method of control may cause a relatively sudden change in piston velocity at the beginning of cushioning action. This could affect machine operation adversely by creating high shock pressures and vibration. For a hydraulic circuit the cushioning action obtainable with a relief valve circuit is more moderate at the initial portion of the cushioning stroke, but increases later.

ROD SEALS

The functions of a sealing arrangement at the rod end of a fluid power cylinder are to minimize leakage of fluid from the cylinder and to exclude foreign materials from entering the cylinder during the return stroke of the piston rod.

Many different types of seals have evolved to cope with the problem of loss of fluid. In the late 1930s, a packing with a "V" shape was developed to control this leakage. This type of packing, commonly used in hydraulic cylinders, worked relatively well. There was little or no collaring of fluid off the rod, and relatively little dripping. However, this type of packing was relatively ineffective in excluding foreign materials from the cylinder. Metallic particles, chips, and other foreign materials adhering to the piston rod were easily drawn past the packing and bearing and into the cylinder. Bearing damage, rod damage, contamination of the fluid, corrosion of parts, and clogging of valves and lines often resulted. Decreased packing life and added maintenance for filter and separator elements occurred.

To cope with the problem of excluding foreign materials, cylinder manufacturers added rod wipers to the gland of their components (Figure 15.19a). However, this often resulted in the oil film being collared off the rod by the wiper and the formation of oil puddles below the glands of cylinders so equipped.

Wipers have undergone considerable development to the extent that, today, a variety of efficient exclusion devices are available. Where materials adhere tightly to reciprocating rods, metal scrapers are often employed. One seal design combines a synthetic wiper and metal scraper in a single unit.

A gland sealing arrangement, developed by Parker Hannifin Company is shown in Figure 15.19b. This assembly employs a primary inner seal to accomplish the major fluid retention job. Because the primary seal is deformed slightly by the pressure of the hydraulic fluid it is sealing (especially in cases where a meter-out circuit is used), some fluid does get past it. This fluid is sheared off the rod by the inwardly projecting lip of a second, outer seal and stored temporarily in a cavity formed by the lip. Since the outer seal is not subjected to fluid pressure, it is not deformed and hence does not leak.

An outwardly projecting lip on the outer seal performs the wiping function to prevent the entrance of foreign material into the cylinder. During the return stroke of the piston rod, the oil stored in the cavity of the inwardly projecting lip of the outer seal is deposited on the rod and carried back into the cylinder past the primary seal and rod bearing. The entire sealing assembly, in the form of a removable cartridge gland, is piloted and threaded to the cylinder head. The assembly can be removed without dismantling the cylinder, and either the seals or the entire assembly replaced.

Figure 15.20 is an exaggerated view of what can happen when a cylinder is too light for the job on which it is applied. An 8, 10, 12, or 14-in. cylinder subjected to excessive

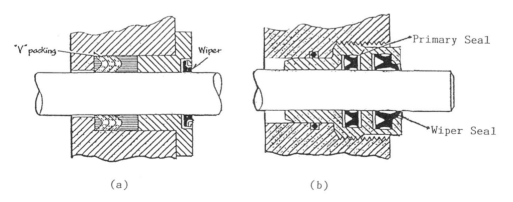

(a) (b)

Figure 15.19 Combination rod sealing arrangements perform dual functions. (a) A lip-type packing reduces leakage of pressurized fluid from the cylinder. A wiper prevents foreign materials from being drawn into cylinder during the return stroke of piston rod. (b) Two synthetic-type seals in an assembly perform a primary sealing function. Oil getting past primary seal is trapped in cavity formed by inner lip of the outer seal and deposited on the rod for return to the cylinder during return stroke. The outer lip of the outer seal performs a wiping function to exclude foreign materials.

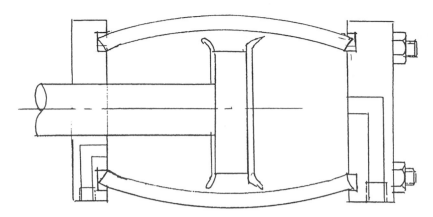

Figure 15.20   An exaggerated view of what can happen when excessive pressure is applied to a large cylinder.

pressure above the designed and rated value coupled with excess torque on the tie rods can permit a change in bore size great enough that the seal on the piston may not stay in contact with the cylinder bore; this is especially true if the circuit is a meter-out where the pressure difference across the piston seal is virtually zero. Seals leak and cylinder screams as it moves back and forth.

The temporary cure for a "singing cylinder" depending on the magnitude of the ailment, may be to put tape (electricians or similar) under the seal I.D. (i.e., fasten the tape to the clean piston surface). This brings the seal diameter up so that it may seal on the O.D. Also for a quick fix, the seal may be stretched for a larger circumference. Backing-off the tie rod torque may help temporarily. The obvious long term answer is to install a cylinder with correct rating!

16
Fluid Motors

The chart in Figure 16.1 provides a quick system test for hydraulic (fluid) motors. A fluid motor may be thought of as an inverse of a pump, but the operation can be significantly different. A check valve-type piston pump will not function as a fluid motor. Certain pumps are designed to be capable of a motor function and certain motors are designed to provide a pumping function in a specific hydraulic circuit. As an example a drag line, anchor windlass, or crane requires power to lift the load (Figure 16.2). As the load is lowered the motor serves as a pump controlling the rate at which the load is lowered. The power in the lowering mode is provided by the drum which in this function is driving the motor/pump. A through shaft may be provided with a parking brake assembled to the motor (Figure 16.3). The brake is applied by the action of a mechanical spring and is released by application of system pressure of a predetermined value. Considering the potential of the shaft breaking under an exceptional load some safety advocates demand that the brake be placed directly on the driven shaft beyond the motor/pump drive coupling (Figure 16.4).

SHUTTLE VALVE RELEASE OF THE BRAKE

A shuttle valve is integrated into a valve assembly to accept a signal from supply line A or B to release a parking brake (Figure 16.5). A counterbalance valve is provided in each motor line. Pressure must be directed to the appropriate motor line to provide the desired force and motion. This pressure can pass through the integral check in the counterbalance valve to meet the resistance created by the motor. The resistance will reflect to the brake release shuttle valve and the counterbalance valve in the opposite motor line. At the predetermined pressure the brake will be released. The load can move at any pressure greater than the set pressure required to release the brake and to open the counterbalance valve holding the fluid captive between the motor outlet and the counterbalance valve.

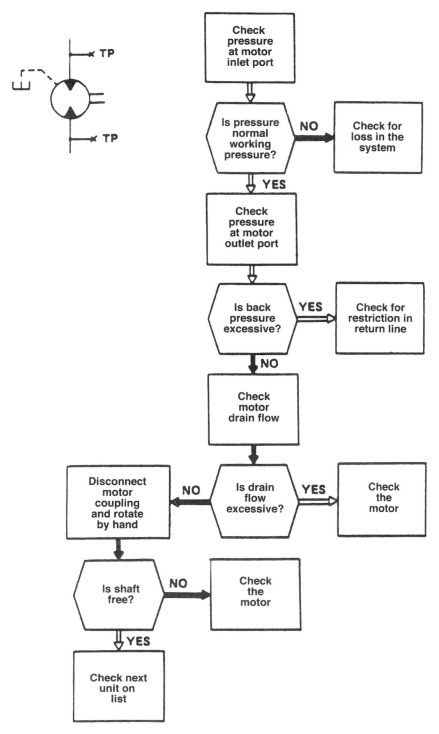

Figure 16.1   System test for hydraulic motors. Courtesy of Vickers, Incorporated.

Figure 16.2 Hydraulic motors may be required to serve as a pump when a load is being lowered to control rate of movement or as a motor when the load is to be lifted in applications such as cranes, drag lines and anchor winches.

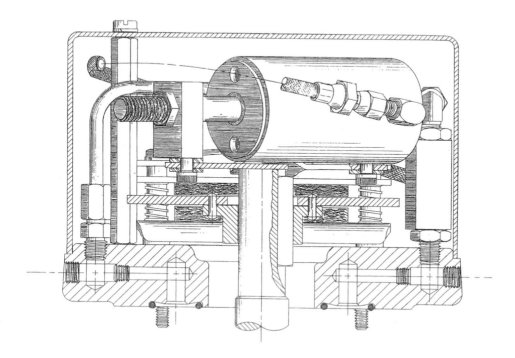

Figure 16.3 A disc brake assembly may be manifolded to the hydraulic motor housing with a shaft extension keyed to the disc for parking functions. This is an on-off function not intended for dynamic braking. Courtesy of Hydraulic Products, Inc.

Figure 16.4   Parking brake assembly may be installed between the motor coupling and the load as a safety feature. Courtesy of Hydraulic Products, Inc.

ANTICAVITATION CHECK VALVES

The valve assembly of Figure 16.6 adds check valves to avoid cavitation created by long motor lines, unusual loads, motor leakage through the housing drain, and safety cross-over relief valves that may be installed within the circuit. An external tank line is optional; its use may be predicated on the amount of captive fluid in the motor circuit and whether a time lag could occur as the counterbalance valve opens to provide oil to the momentary cavitational activity. A line to the tank provides oil immediately to satisfy the momentary vacuum developed in the circuit.

BASIC MOTOR CHARACTERISTICS

Basic motor characteristics must be understood to troubleshoot each motor type. For example, a gear pump is a rugged, hard working source of pressurized fluid, but as a motor the gear structure has some limitations. The gear motor is considered best suited for easily started loads such as fan drives and certain types of traction drives that develop into high horsepower loads when up to speed. Gear motors are often applied to speed reduction units for heavy rotary force and motion tasks such as boring holes for telephone poles.

Centrifugal force causes pump vanes to follow the internal cam surface. Vane motors

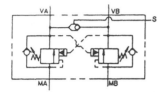

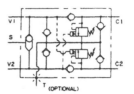

Figure 16.5 A shuttle valve located up-stream of the counterbalance valves can serve to provide pressurized oil to release the parking brake.

Figure 16.6 Anticavitation check valves can be integrated into the motor circuit in addition to the shuttle valve if long lines or unusual loads can create a momentary negative pressure value.

must have a suitable mechanism to be certain the vanes are in position in the stalled or starting position because centrifugal force is not a part of the motor function at the starting mode. A spring or springs may be employed to mechanically push the vane toward the camring (Figure 16.7). Other methods include pressurization of the vane slot or mechanical pins with a cam structure.

AXIAL AND RADIAL PISTON MOTORS

Bent axis and cam plate (or axial piston) motors employ a pair of kidney-shaped ports to direct fluid to and from the pistons. The bent axis motor is provided with multiple bearings for axial and radial loads to start heavy loads (Figure 16.8). The cam plate (axial piston) motor shares these capabilities (Figure 16.9). These motors can also be provided as variable displacement units with a limited speed change by modifying the piston stroke. Certain limitations can be anticipated because of the inherent drive angle of the piston assembly. At the greatest angle the maximum torque is provided; as the angle is reduced the rotating speed increases and the torque value decreases. Some models have internal stop devices to insure a proper operating range. Tampering with this mechanism can cause the motor to exceed the rated speed and/or lose torque. A wide speed range can be accommodated by bent axis and other axial piston motors.

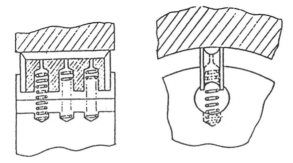

Figure 16.7 Compression springs may be used to position motor vanes.

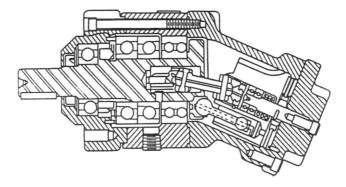

Figure 16.8   Bent axis motor functions equally well as a pump.

THE COMMUTATOR VALVE

Radial piston motors with a commutator valve to direct the pressurized flow to selected pistons can provide good stalled torque characteristics (Figure 16.10). One type of commutator mechanism is called a pintle valve. The pintle valve employs a rotary valving member within a ported sleeve (Figure 16.10). These radial piston motors, available with a commutator valve for high-torque low-speed operation, offer some significant advantages in space required and loads that can be moved. The radial piston motor may require counterbalance valves to prevent the motor from functioning as an open suction pump. The connecting rods on some designs are not capable of retracting the pistons in an uncontrolled pumping mode with resulting damage to the piston rod assembly. The counterbalance valve restricts flow from the motor so that the piston assembly is suitably restrained and the motor can decelerate the load at the rate determined by the supply to the motor inlet. When troubleshooting high-torque low-speed piston motors with commutator valving make certain that the counterbalance valve setting is greater than the maximum induced load that the motor will encounter.

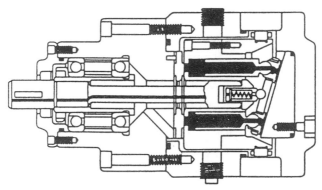

FIXED DELIVERY (CAM PLATE)

Figure 16.9   Cam plate-type axial-piston motor.

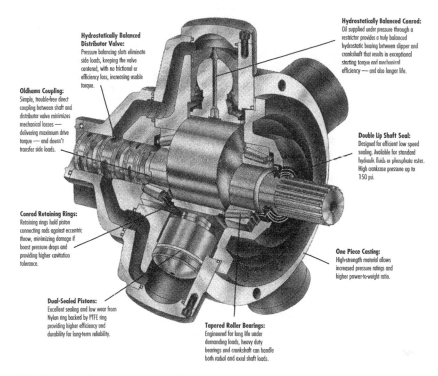

Figure 16.10 — image labels:

Hydrostatically Balanced Distributor Valve: Pressure balancing slots eliminate side loads, keeping the valve centered, with no frictional or efficiency loss, increasing usable torque.

Hydrostatically Balanced Conrod: Oil supplied under pressure through a restrictor provides a truly balanced hydrostatic bearing between slipper and crankshaft that results in exceptional starting torque and mechanical efficiency — and also longer life.

Oldhams Coupling: Simple, trouble-free direct coupling between shaft and distributor valve minimizes mechanical losses — delivering maximum drive torque — and doesn't transfer side loads.

Double Lip Shaft Seal: Designed for efficient low speed sealing. Available for standard hydraulic fluids or phosphate ester. High crankcase pressure up to 150 psi.

Conrod Retaining Rings: Retaining rings hold piston connecting rods against eccentric throw, minimizing damage if boost pressure drops and providing higher cavitation tolerance.

One Piece Casting: High-strength material allows increased pressure ratings and higher power-to-weight ratio.

Dual-Sealed Pistons: Excellent sealing and low wear from Nylon ring backed by PTFE ring providing higher efficiency and durability for long-term reliability.

Tapered Roller Bearings: Engineered for long life under demanding loads, heavy duty bearings and crankshaft can handle both radial and axial shaft loads.

Figure 16.10 Radial piston motor with pintle-type commutator valve. Courtesy of Precision Machinery Division, Kawasaki Motors Corporation, U.S.A.

THE ORBIT MOTOR

A Gerotor-type pump functioning as an internal gear pump or motor typically creates a series of pockets that can be filled from a contoured cavity much like the axial and bent axis pump. The inner pinion and outer ring gear rotate (Figure 16.11) to provide the displacement function.

Orbiting-type motors provide high torque in a small package by using commutator valves to direct fluid to and from the cavities within the Gerotor-type displacement set.

The basic displacement element (Figure 16.12) is generated rotor-type assembly. It is not intended to rotate as in the design of Figure 16.11. The outer ring-gear is held stationary in the assembly of Figure 16.13. Note that the inner lobes of the ring gear are formed by use of solid or tubular rolls as an alternate design to that of Figure 16.12. The inner pinion is free to move in an orbital path as pressurized fluid is directed to selected pockets created by the rotor set and exhaust fluid is directed to a lower pressure area from diametrically opposite pockets. The orbiting pinion is connected through a shaft assembly that serves as a universal joint to the output shaft. An extension of this universal joint at the opposite end of the motor drives the lapped flat commutator valve assembly.

The commutator employed in the motor of Figure 16.14 is of the pintle-type directing fluid in the required pattern to the orbiting pinion in the displacement set. Models are available with dual displacement sets that can be selectively valved to provide maximum torque or an optional increase in rotating speed with a lesser torque value. The valving

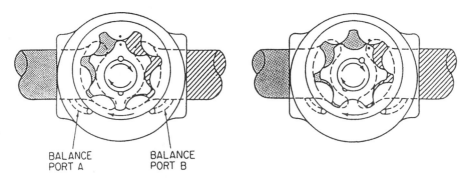

BALANCE   BALANCE
PORT A   PORT B

Figure 16.11 Gerotor-type hydraulic pump/motor. Ring gear is housed eccentric to pinion and drive or driven shaft. Courtesy of Fluid Power Division Parker Hannifin Corporation.

function is created by an orbiting ring member in the motor of Figure 16.15. A through shaft can be provided with this design.

The valving function of the motor of Figure 16.16 is driven by a universal drive shaft from one side of the orbiting pinion. Power is transmitted to the output shaft by a universal joint at the other side of the pinion. The motor of Figure 16.17 is fitted with a valving assembly that can selectively bypass the appropriate cavities in the rotor set so that a higher speed is available with lesser torque. When fluid is directed to all of the cavities in the normal pattern the torque is increased and the speed is decreased in the same proportion.

HYDRAULIC MOTORS MAY BE INTEGRATED INTO WHEELS OR DRUMS

The basic fluid motor structure may become an integral part of a traction drive. Wheel motors, winch drum motors, and comparable installations integrate the motor structure into the basic force and motion device to reduce parts, minimize piping, and save space and at times weight.

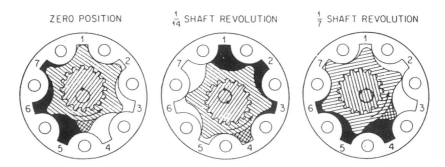

ZERO POSITION $\frac{1}{14}$ SHAFT REVOLUTION $\frac{1}{7}$ SHAFT REVOLUTION

Figure 16.12 Gerotor-type displacement element designed for orbit service. Commutator valving directs fluid to rotor set pockets much like the commutator employed in an electric motor. Courtesy of Fluid Power Division Parker Hannifin Corporation.

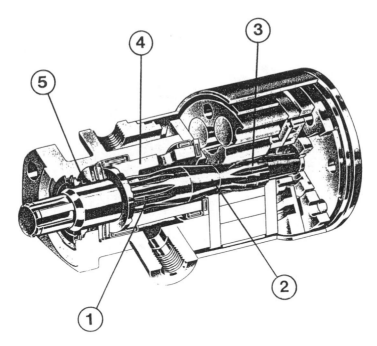

Figure 16.13 Orbit-type motor. (1) Spline drive to output shaft. (2) Drive shaft serving as universal joint. (3) Spline drive within orbiting pinion with extension to actuate face-type commutator valve. (4) Output shaft and bearing assembly. (5) Shaft seal. Courtesy of TRW Ross Gear Division.

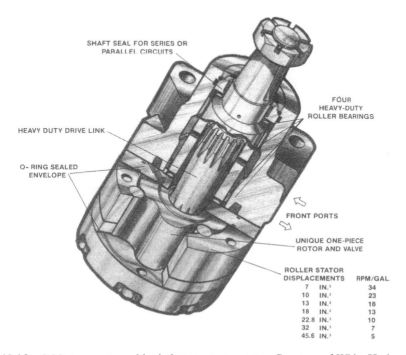

Figure 16.14 Orbit-type motor with pintle-type commutator. Courtesy of White Hydraulics, Inc.

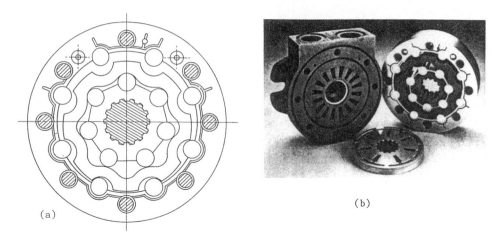

Figure 16.15 Orbit-type motor with internally generated rotor structure. Central ring gear orbits between pinion on shaft and inner surface of outer fixed housing ring. (a) Displacement assembly. (b) View of valving mechanism. Courtesy of Nichols Fluid Power Division, Parker Hannifin Corporation.

TROUBLES AND PROBABLE CAUSES

1. Scoring: Cylinder block and face, bores, valve plate, and pistons. Probable cause: Fluid contamination.
2. Speed drop-off and heat. Probable cause: High cross-port leakage resulting from fluid contamination affecting cylinder block/valve plate surface finish.
3. Main bearing failure. Probable cause: Coupling misalignment; shaft seal leakage washing lubrication grease out resulting in bearings running dry and failing.
4. Blown shaft seal. Probable cause: High case drain pressure.

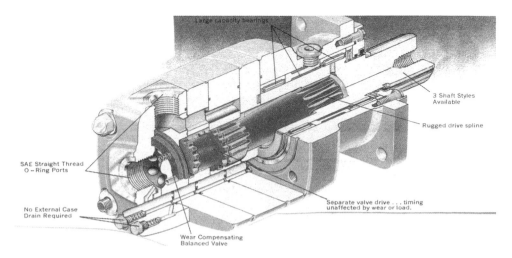

Figure 16.16 Orbit-type motor with wear compensating balanced valve. A separate valve drive isolates the load so that timing is not affected by load or wear. Courtesy of Eaton Corporation Fluid Power Operations.

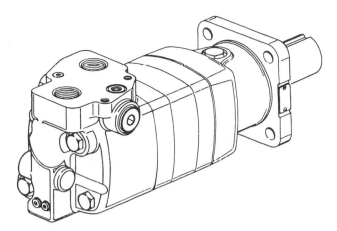

Figure 16.17   Orbit-type motor with valving to provide low speed and full torque with one position of the valve. The second position of the valve provides double speed and one-half torque. Courtesy of Eaton Corporation Fluid Power Operations.

17

Energy Storage, Shock Suppression, and Supplemental Fluid Flow:
Tasks for the Hydraulic Accumulator

The two most frequently used accumulator types are the bladder (Figure 17.1) and piston designs (Figure 17.2). Accumulators are used for various functions in hydraulic systems, including energy storage, pulsation dampening, thermal expansion, and surge control.

This chapter discusses accumulators that are used primarily for energy storage. Several operating parameters must be addressed when sizing accumulators, including:

1. Maximum and minimum system operating pressures as required by system performance.
2. Fluid discharge that meets predetermined flow requirements.
3. Selecting a gas (nitrogen) prefill pressure that satisfies 1 and 2 above.

The preferred mounting positions are vertical with the discharge (fluid) port down and the gas connection up. Under some circumstances the accumulator may be mounted horizontal.

SPRING-BIASED ACCUMULATOR

Storing energy can also be accomplished with a spring-biased piston assembly (Figure 17.3). A spring-biased accumulator may be integrated into the design of special machines to level out shock conditions and/or supply relatively small quantities of fluid at a high flow rate. Major service requirements for spring-biased accumulators result from contamination damaging the piston, seal, or bore. Repair usually entails cleaning and replacing piston seals. Scoring of cylinder bore may require remachining or replacement.

WEIGHTED ACCUMULATOR

A vertical cylinder assembly with weight added to provide gravity forces to develop pressure on the fluid provides an accumulator function with significantly different characteristics than the gas-biased or spring-biased designs (Figure 17.4).

Gas-charged accumulators respond to the gas laws. Spring-biased accumulators

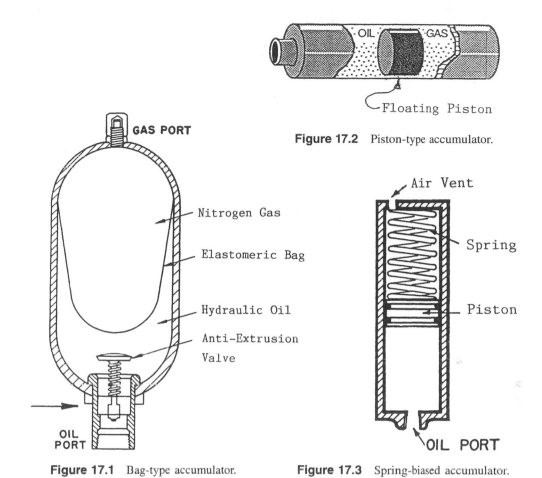

Figure 17.2 Piston-type accumulator.

Figure 17.1 Bag-type accumulator.

Figure 17.3 Spring-biased accumulator.

relate to the pounds per inch of compression of the bias spring and the installed characteristics. The weighted accumulator provides a uniform line pressure from initial entry of fluid until the piston reaches maximum operating stroke. Control is usually by means of a limit switch at the maximum stroke to stop flow to the accumulator and a switch at the lower discharge limit to resume supply flow to recharge the accumulator. Typical use for the weighted accumulator is to maintain a pressure for a molding press, rolling mill pressure, counterbalancing large weights, and other applications where pressure must be maintained at a fixed predetermined level. The pressure value of a weighted accumulator can be changed by increasing or decreasing the weight on the cylinder assembly.

Troubleshooting may involve damage from contaminated fluid, worn packing, or incorrectly adjusted limit switches. Problems may also be involved in other parts of the hydraulic circuit. The weighted accumulator will not absorb shock loads. It is more likely to create shock loads because of the rigid weight to area relationship. The weighted accumulator usually has a high weight-to-area relationship so that gradual acceleration and deceleration within the circuit is essential when using this device to hold a steady pressure on the machine member(s).

Weight Calculated for
Desired Pressure

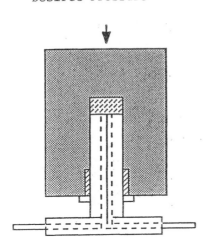

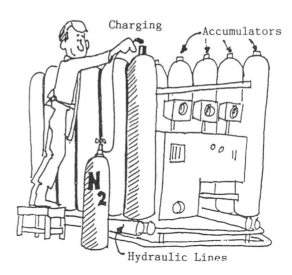

Charging   Accumulators

N 2

Hydraulic Lines

Figure 17.4 Weighted accumulator.

Figure 17.5 A bank of accumulators may be employed when additional capacity is needed.

MULTIPLE ACCUMULATORS IN BANKS

Accumulators may be assembled in multiples to provide the needed capacity. This is usually much more economical than designing and building a larger capacity pressure vessel. The multiple assembly mode can be much easier to service (Figure 17.5).

A SUPPLEMENT TO PUMP CAPACITY

Accumulators supplement system pump capacities. They provide an additional supply of pressurized fluid for immediate use. Being "on line" they react instantly to system pressure drops. They deliver fluid to the system when system pressure goes below the pressure level of the gas pressure in the accumulator. Conversely, they will accept fluid available from the system up to the setting of the system relief valve or the setting of the compensator of a variable-displacement pressure-compensated pump. These characteristics frequently allow the system to use a smaller pump than would otherwise be required.

A typical accumulator application when using a pump that must be unloaded to the system reservoir between operating cycles follows. The stored fluid under pressure in the accumulator flows out into the system at times of peak demand or at any time when pump capacity is insufficient to maintain system demand within predetermined pressure ranges. At that time in the operating cycle when zero flow is required (usually at the end of an operating cycle) pump flow is available to recharge the accumulator to the maximum desired system pressure. At the completion of the recharging operation the pump is then unloaded to tank.

UNLOADING THE PUMP

Two popular means to unload the pump are:

1. A dual setting pressure switch can provide an electrical signal at either maximum or minimum system pressure. When activated by maximum system pressure, the electrical signal from the pressure switch deenergizes a solenoid operated venting valve; this allows the system relief valve to unload the pump to tank (Figure 17.6). When system pressure drops to the minimum pressure level, the venting valve is energized by the pressure switch signal; this causes the relief valve to close thus directing pump flow back into the system.

2. An all-hydraulically operated unloading valve can perform the same function. These valves usually have a fixed pressure differential value between the two operating pressures (Figure 17.7). The unloading piston is slightly greater in diameter than the pilot poppet. As the pilot poppet opens it unbalances the unloading piston; this provides a snap-action to completely unload the main relief valve poppet. The accumulator holds the pilot poppet off the seat by the action of the unloading piston. As pressure decays in the accumulator circuit the pilot poppet reseats and forces the unloading piston away from the pilot poppet. The pump then charges the accumulator system and the process will be repeated.

THE TIME FACTOR

When pump capacity is available to recharge the accumulator to its full fluid condition there is a time element required. Each system has its own repetitive rhythm such as recharge time, actuator motions, and pump and control sounds. A change in these rhythms assists the troubleshooter in selecting the correct direction to take in searching out the malfunction.

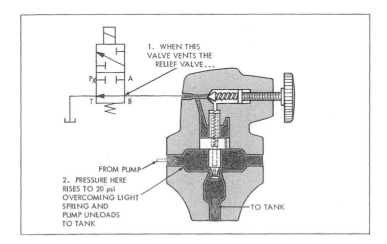

Figure 17.6 A solenoid-operated pilot valve can be used to vent a relief valve. A check valve in the accumulator supply line prevents loss of system pressure when relief valve is vented. Courtesy of Vickers, Incorporated.

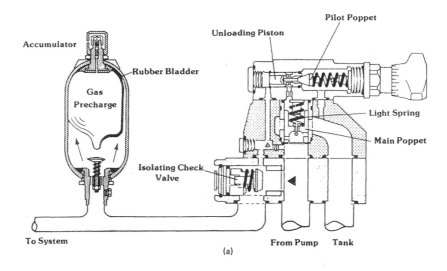

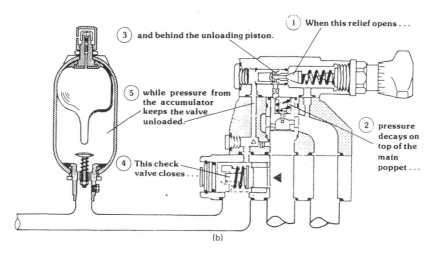

Figure 17.7 A relief valve assembly designed with an unloading piston and isolating check valve to prevent loss of system pressure when relief valve is vented. (a) Charging the accumulator. (b) Relieving and unloading of the pump. Courtesy of Rexroth Corporation.

The use of human senses of sight, sound, and feel assist the troubleshooter during the malfunction resolution process, so think first before troubleshooting.

TROUBLESHOOTING

Bladder-Type Accumulator

Problem: Accumulator Bladder Ruptures

When a bladder rupture occurs the normal time to recharge the accumulator is completely changed. This is also true of the unloading function associated with controlling pump flow either to tank or into the system.

As the prefill gas supply works its way out of the accumulator through the system and back to tank, what was a relatively smooth and controlled pump unloading sequence turns into an erratic and noisy series of events.

The time to reach maximum system pressure is nearly instantaneous. At maximum pressure the unloading system redirects pump flow to tank. System pressure quickly drops to minimum and the unloading system reverses pump flow back into the system. The result is rapid and repetitive unloading system reaction with accompanying piping vibration and control valve off/on noise. These abnormal fluctuations tell the troubleshooter what the malfunction is.

Action

Replace the bladder; follow the manufacturer's instructions.

When bladder rupture occurs it is quite common for the bladder to be compacted into the top of the accumulator shell. A fast method of detecting bladder rupture is to depress the prefill valve stem. Lack of gas pressure equates to very little force required to depress the valve stem. Also, a small oil seepage usually occurs. *Caution*: If the valve stem does not move easily and if no oil appears, then revert to using the precharge assembly with pressure gauge to test for gas prefill pressure.

Problem: Partial Loss of Gas Prefill Volume

If a leak develops at the gas prefill valve, the result will be a deterioration of system performance due to decreased discharge capacity from the accumulator.

Example: A system requires approximately 280 in.^{3} of fluid from the accumulator. A 5-gal size accumulator is being used. The maximum system pressure is 2500 PSI and minimum is 1500 PSI. A gas-filled pressure of approximately 1100 PSI is needed to provide for 280 in.^{3} cubic inches of fluid discharge within the maximum and minimum system pressure.

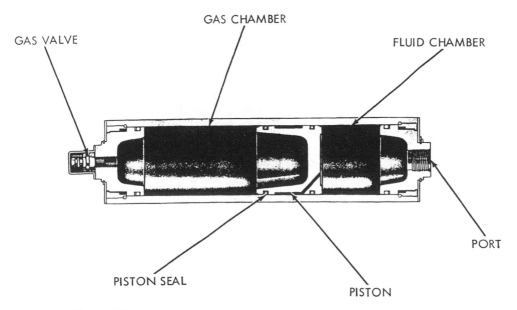

Figure 17.8 Piston-type accumulator. Courtesy of Vickers, Incorporated.

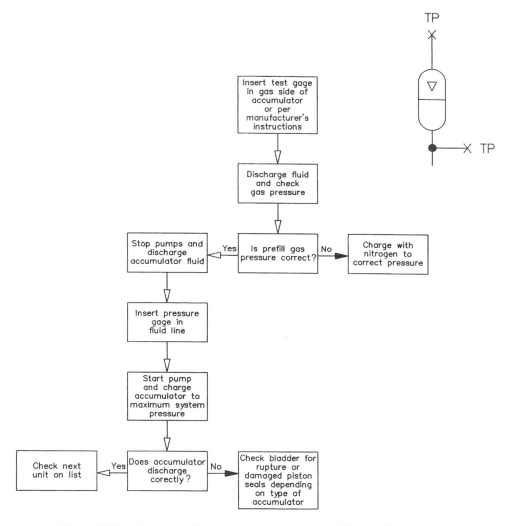

Figure 17.9 System test for accumulators. Courtesy of Vickers, Incorporated.

If 500 PSI of gas prefill is lost, then the available discharge of oil drops to approximately 190 in.^{3}. The result will be system actuator performance deterioration. If the gas prefill pressure loss is slow then system performance degradation will also be slow. Troubleshooting for system performance degradation requires careful analysis. Systems having accumulators should be a tip off to the troubleshooter to check the accumulator prefill pressure first if no other obvious signs are in evidence.

Action

Determine the cause for the leak and repair or replace the bladder as required. Follow the manufacturer's instructions.

Piston-Type Accumulators

The piston-type accumulator incorporates a free-floating piston separating the nitrogen gas chamber and the system pressurized oil. At one end of the cylinder a gas (e.g., nitrogen) prefill valve is provided to allow filling with gas. The port for passage of fluid to and from the system is located at the other end of the piston-type accumulator (Figure 7.8).

Malfunctions with piston accumulators usually are related to fluid cleanliness since contaminants affect the positive seal required by the piston sealing element.

Problem: Loss of Gas Prefill Pressure

The operational result of the loss of prefill gas either from gas leakage at the prefill valve or past the piston seals results in system performance deterioration in a similar manner to the bladder-type accumulator.

Action

Check for loss of gas prefill pressure. If the loss is through the prefill valve repair or replace it. If it is not the valve or the seal related to the accumulator barrel and head cap assembly, then it is leakage past the accumulator piston. Check the piston seals and the cylinder barrel bore for wear or scoring. If the barrel is worn or scored beyond repair then it must be replaced. Also check the condition of the piston and piston seals.

SYSTEM TESTS FOR ACCUMULATORS

The chart of Figure 17.9 provides a quick reference to check the accumulator installation.

18
Heat Exchangers

HYDRAULIC COOLERS

Two popular styles of heat exchanger commonly referred to as "coolers" used in hydraulic systems are: 1) the water type (Figure 18.1), and 2) the air type (Figure 18.2). The purpose for coolers is to maintain an acceptable reservoir oil temperature of approximately 120–125°F.

Coolers supplement the cooling capacity of the reservoir, which is in itself a heat exchanger. When ambient air temperatures are lower than the reservoir oil temperature, heat is dissipated through the outside surfaces of the reservoir.

Proper sizing of coolers requires an analysis of the heat generation characteristics of the hydraulic components in the system. Heat generation results from mechanical friction and internal leakage in hydraulic system components.

Working clearances in components represent orifices through which system fluids flow under pressure, resulting in internal leakage. The energy being used for useful work is converted into heat resulting from the pressure drop that takes place during the periods of internal leakage.

Cooler malfunctions are attributable to the working condition of the cooler or a change in volume by temperature of the cooling medium. A bypass relief valve piped in parallel to the cooler is recommended for protection against system oil flow surges or pressure spikes. An "in-line" check valve is an effective bypass relief valve.

TROUBLESHOOTING AIR-TYPE COOLERS

Problem: Loss of cooling efficiency

1. Air coolers of the automotive radiator design require clean air passages. An unacceptable rise in outlet oil temperature could signal the need for an inspection and cleaning of the air passages. A visual inspection is usually sufficient to confirm that this is the problem.
2. If the cooler has a bypass relief valve, a check should be made to determine if the

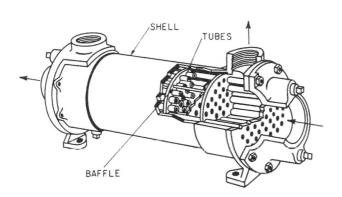

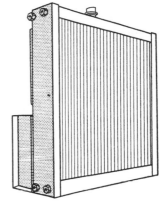

Figure 18.1 Shell and tube heat exchanger: oil-to-liquid heat transfer.

Figure 18.2 Honeycomb radiator-type heat exchanger: oil-to-air heat transfer.

bypass valve is open, thus allowing system fluid to bypass the cooler. The problem could be a partially contaminated and plugged cooler or a contaminated bypass valve that will not allow full system volume to pass through the cooler.

Problem: System fluid leakage

1. Using a pressure gauge, check pressure at the inlet to the cooler for excessive pressure or pressure spikes. Follow the manufacturer's recommendations for approved cooler oil pressure.
2. If the system has a bypass relief valve, check it for proper operation.

TROUBLESHOOTING WATER-TYPE COOLERS

Problem: Loss of cooling efficiency (Figure 18.3)

1. Check both the temperature and volume of the cooling water. A deficiency in either or both of these factors will contribute to excessive cooler outlet oil temperature. Take corrective action for either deficiency.
2. Check for proper operation of the fluid bypass relief valve if one is used. A quick sensory check can be made by feeling the piping by hand. If the system fluid temperature downstream of the bypass valve is similar to that upstream of the valve, then fluid is probably flowing through the valve.
3. If the cooling system has an oil temperature control system, then it must allow adequate water flow through the cooler to maintain desired oil temperatures (Figure 18.4). Effective water temperatures are still a requirement.

Problem: System fluid leakage

1. Cooling water passes through tubes inside of the cooler. System fluid is contained in the shell of the cooler. If an internal water leak develops, water will contaminate the entire hydraulic system. Water in system fluids changes the color and viscosity of the fluid.

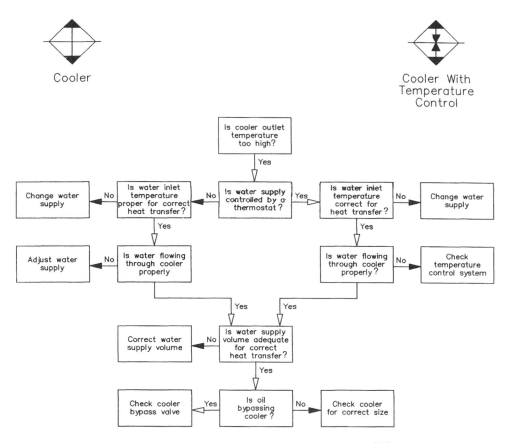

Figure 18.3   System test for oil-to-liquid heat exchanger. Courtesy of Vickers, Incorporated.

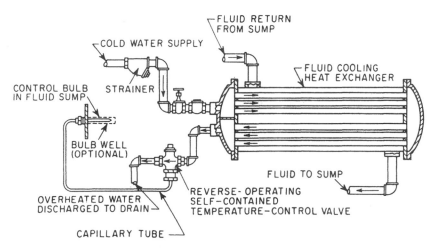

Figure 18.4   Oil-to-liquid heat exchanger fitted with an automatic temperature control.

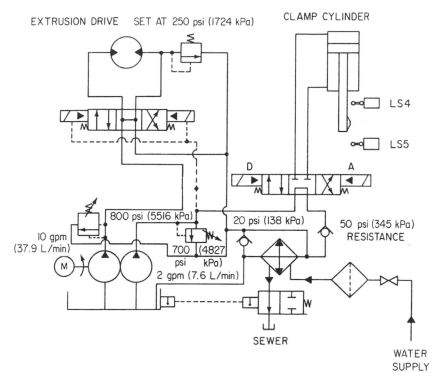

Figure 18.5 Typical extrusion machine hydraulic circuit with oil-to-water heat exchanger and automatic control for cooling water flow.

2. Water also affects the lubricity of the fluid, which can result in severe damage to system components, especially those that rotate, such as hydraulic pumps and motors.
3. Once water has contaminated the system fluid supply it must be completely removed and replaced with new, uncontaminated fluid. It is good practice to flush the system with new fluid as part of the fluid replacement process.

A TYPICAL APPLICATION

A plastic extrusion press hydraulic system equipped with an oil-to-water heat exchange unit is illustrated in Figure 18.5. Note the filter symbol shown ahead of the water cooler. The filter is essential where raw water is taken from a local source to be used in the heat exchange process. Some filters are quite coarse with the intent to keep out beer cans, frogs, fish, and major contaminants. Other filtering systems can be quite sophisticated to prevent silting of the cooler structure with finer contaminants. The water control valve is shown as a thermally operated unit. A thermal electrical switch could be used in place of the capillary control bulb within the sensing area of the reservoir. A solenoid operated water valve can be used to open and close the water valve when a thermal temperature switch is employed for sensing the temperature within the reservoir.

19

Preventive Maintenance

Why learn the value of preventive maintenance by accident? Avoid the costly and unpleasant surprises of emergency repairs. Extend equipment life. Reduce equipment downtime. All it takes is a preventive maintenance program that includes careful planning, good workmanship, and a workable record system.

PREVENTIVE MAINTENANCE CHECKLISTS

Equipment operators can help make your preventive maintenance program a success, if you provide checklists that spell out maintenance requirements and responsibilities (Figure 19.1). Develop your own maintenance checklist from manufacturer's literature on your machines and their hydraulic components. The check list can be any size, from a wallet card to a wall chart or both. It should be used with equipment control sheets, vehicle reports, and maintenance records, all of which should be filed for ready reference. Keep the checklist plain and simple so that it's easy to understand and use. Both manufacturing machinery and mobile equipment can benefit from good maintenance records. Many companies keep the records in computer programs that make data immediately available to call-up screens at remote stations.

OPERATION UPDATE

Encourage equipment operators to develop and maintain a good working knowledge of the hydraulic systems. Encourage them to seek certification as a fluid power mechanic through the Fluid Power Society certification programs. Help them "keep current" by passing along helpful literature, arranging for periodic programs on equipment maintenance and encourage their affiliation with educational groups such as the Fluid Power Society and others. Local community colleges and technical high schools often provide both seminars and short courses covering fluid power equipment and systems and general technology courses. Such sessions and society meetings, developed with the help of oil suppliers or equipment

Figure 19.1 A well thought out and realistic checklist makes the equipment operator a key player in the preventive maintenance program. Courtesy of Commercial Intertech.

manufacturers, usually include an update on good hydraulic troubleshooting fundamentals and repair techniques.

THE OPERATOR'S JOB IN PREVENTIVE MAINTENANCE

Spell out just what the operator should do for proper equipment care, such as routine checks and records, or checking oil level and condition, removing excess dirt from the equipment, turning in equipment for oil changes and maintenance, and reporting any loose, damaged, or deteriorating parts. Keep operators conscious of the need for continuing inspection including: reporting worn or damaged tubing and hose, foaming or turbulence in the oil reservoir, signs of overheating, oil leaks, loose joints, worn seals, and any unusual noises that might mean cavitation, trapped air, or worn bearings and seals. Encourage the operator to note and report any slow down of machine operating time that might be an early warning of future problems.

THE MAINTENANCE MAN

It is up to the maintenance man, more than anyone else, to make the preventive maintenance program work; these items should certainly be on the "Do" list:

1. Follow the manufacturer's recommendations on oil grade and viscosity, filter replacement, system purging, and periodic flushing.
2. Make periodic checks of the equipment to make certain it is functioning properly. Know the condition of all equipment, at all times, and schedule minor or major maintenance accordingly.
3. Disassemble or flush clean the entire hydraulic system when replacing a component that has failed.
4. Make certain that replacement components or parts have been stored properly to prevent contamination (Figure 19.2). Oil drums should be stored in racks under cover with proper identification. Facilities to move oil from the drum to the machine reservoir can be a key item in minimizing downtime.

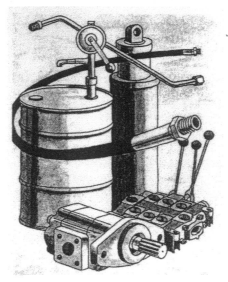

Figure 19.2   Proper storage of repair parts and service materials can help prevent contamination of the fluid power system. Courtesy of Commercial Intertech.

PROTECTING THE LIFE BLOOD OF THE MACHINE

Hydraulic reservoirs are relatively large for open circuit hydraulic systems in which the oil circulates from the pump outlet through the circuit and then back into the reservoir. Hydraulic reservoirs for closed circuit hydraulic systems may be very small. In a closed circuit the oil circulates from the pump outlet to an actuator and from the outlet of the actuator back to the opposite port of the pump. The pump serves as the pressure control, directional control, and flow control. It is reversible, usually pressure compensated, and capable of infinite position in either direction of flow to provide the desired rate of actuator movement. The reservoir may be within the housing of the pump or the housing of the pump and actuator combined. A relatively small charge pump, usually constant displacement, is provided for a typical closed circuit hydrostatic transmission to circulate oil within the low pressure side of the hydrostatic circuit. In the process the oil may be cooled and filtered. This charge pump function insures solid pressure on the suction side of the pump to make up for lubrication losses in the pump and actuator, provides pilot pressure for controls, provides facilities for filtering the fluid and, usually provides a circulation through an external supplementary cooling system (if needed). The closed circuit hydrostatic assembly may have a very efficient cleaning and cooling system for the oil. It is not subject to many of the hostile environmental areas in which the typical open circuit transmission system must operate.

The widely used open circuit hydraulic system reservoir may be fabricated in many different shapes. They reflect need for the fluid storage on a hydraulically actuated machine and in some designs may become an integral part of the machine. Some are of a standardized, catalogued design; others are custom designed and built to fit into available space on or near the machine. Because the reservoir often serves as a major heat transfer component it may be desirable to locate it where greatest radiation potential can be obtained.

In addition to the heat transfer function the reservoir is expected to protect and store the oil in the normal force and motion function of the hydraulic circuit.

FILLING THE RESERVOIR

The fill pipe is an obvious point for entry of contaminants. Because of this the fill pipe is often fitted with a fine screen. The fill pipe assembly may also be fitted with an air breather that can include an air filter element. The air breather function may be a separate mechanism where large air flows can be expected.

Because the fine screen in a fill pipe will restrict the gravity flow from a container some mechanics cannot resist the temptation to punch a hole through the screen with a screwdriver or other tool to expedite the fill process. This, of course, opens up the potential for major contamination. It is also true that the fill pipe may be in a dirty area and the mechanic may not be meticulous in removing potential contaminants in this area.

Recognizing the difficulty in controlling the fill cycle with a relatively unprotected access system, some designers have virtually eliminated the contamination potential by providing a connection in a convenient hydraulic circuit tank line of appropriate size ahead of the return line filter. To avoid disturbing the piping the connection can be made through a shut-off valve with a protective plug in the external connection to the valve which can be removed when fluid is to be added to the system. Some users have chosen to install one half of a quick-connect coupling at the fill point with a protective dummy cap when the fill function is completed. In addition to the fill facility a second connection can be provided at the lowest point in the reservoir for draining water or other contaminants. A portable filter dolly (Figure 19.3) can be wheeled to the machine, the transfer pump suction connected to the valve at the low point in the reservoir and the filter discharge connection joined to the fill point and then the fluid can be passed through the filter on the dolly. A thorough filtering of the residual oil in the tank can be provided. This procedure can be scheduled on a regular basis when the machine is operating in a hostile environment or on an intermittent basis after a hose or other component has been replaced. Obviously connections must be carefully cleaned to avoid ingestion of contaminants.

Oil from storage may have been contaminated. By transferring the oil from the storage barrel to the machine through the filter dolly, the major filter on the dolly will remove contaminants and the return line filter in the machine hydraulic circuit will not be overloaded with new contaminants.

FILLING WITH THE CORRECT OIL

All fill connections for various machine reservoirs using the same fluid medium should have a common size of fill valve or quick connection. Reservoirs using other fluids should have different sizes of fill valve or quick connection to avoid filling with incorrect fluid.

As an example, all machines using a high quality petroleum-based 200 SSU oil may use a ½-in. connection. Heavier oil for press circuits, as an example, may use a ¾-in. connection; those using a phosphate ester may have a ⅜-in. connection, while water glycol units may use a 1-in. connection. If quick-connect couplings are used it is possible to alternate male and female connectors to differentiate fluid types. Color coding reservoirs,

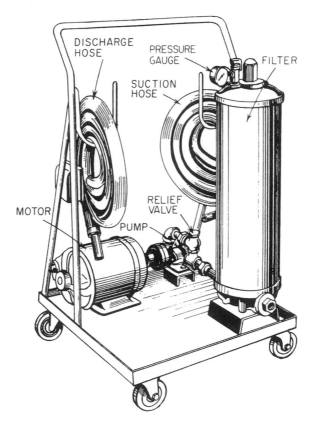

Figure 19.3 Filter dolly includes transfer pump, filter, and interconnecting hose.

filter dolly, and associated lines can also help to insure that the proper type and grade of fluid is added.

CONTAMINATION VIA THE AIR BREATHER

Many types of air breather and filter combinations are available. Service is usually a function of the ambient contamination level in the area of the reservoir.

An alternate option is the use of a pressurized reservoir. An auxiliary air tank may be included to compensate for actuator displacement. In any design sufficient air capacity must be provided above the oil level to compensate for the displacement of the actuators. The pressurized reservoir insures a positive supply of oil to the pump suction. Aircraft hydraulic systems use the pressurized reservoir to compensate for differing air pressures at various altitudes. A breather bag (Figure 19.4) permits entry of air to compensate for rod displacement. The breather bag is fitted internally at the top of the reservoir with the neck open to atmosphere. It is a synthetic rubber bag that provides a permanent flexible nonporous barrier between the atmosphere and the system fluid without affecting the operational functions of the system components. The bag must be of sufficient size to accommodate the volume change within the tank, equal to the full system displacement,

MINIMUM OIL VOLUME IN TANK   NORMAL OIL VOLUME IN TANK   MAXIMUM OIL VOLUME IN TANK

Figure 19.4 Internal breather system. The Fawcett Breather Bag isolates atmospheric contaminants from hydraulic oil and compensates for cylinder rod displacement. Courtesy of Greer Hydraulics.

plus an additional 25% to allow for any leakage and volume change due to thermal expansion or contraction.

A flexible bladder fitted into a shell can provide an isolation means to eliminate moisture and contamination potential from the hydraulic reservoir (Figure 19.5). Greer Hydraulics recommends that every KleenVent installation be equipped with a functional pressure/vacuum relief valve to protect the reservoir in the event of sudden fluid loss. The KleenVent assembly should be installed when the reservoir level is at its high mark. This will allow the bladder to deflate as the reservoir level drops. The unit should be installed in a vertical position for optimum performance. A sealed reservoir requires the use of a fill system such as the filter dolly or its equivalent.

RECORDS

Keep a detailed Equipment File on the rig, to show if a given malfunction has occurred in the past and whether it was corrected. This file should include an Operator's Report and the Shop Inspection Report or a combination of the two. These records, properly kept, will help make your preventive maintenance program a successful one. Avoid unpleasant surprises.

CONTROLLING WEAR

There's no way in the world to completely stop wear on moving parts in machinery or equipment. But you can greatly reduce wear in hydraulic components if you:

1. Make sure the oil is kept in good condition and at the right level.
2. Do your best to keep dirt and water out of the system.
3. Correct any condition that causes heat build-up.

Oil Level

Oil level should be checked *every day* that the equipment is used. Stress the importance of this daily inspection to your operators. Functions that require a lot of oil—extending a

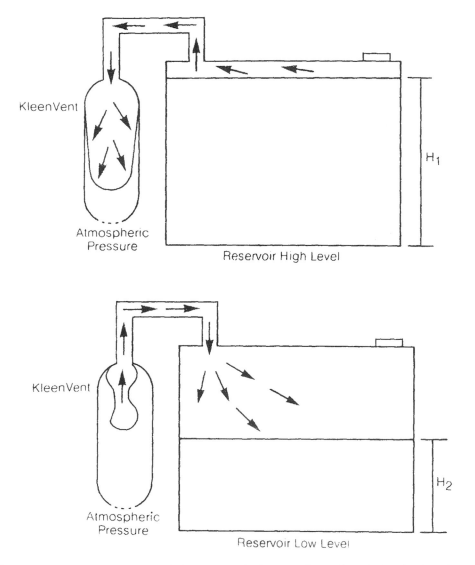

KleenVent

Atmospheric
Pressure

H_1

Reservoir High Level

KleenVent

Atmospheric
Pressure

H_2

Reservoir Low Level

Figure 19.5   An elastomeric bag isolates atmospheric contaminants from hydraulic oil and compensates for cylinder rod displacement. Courtesy of Greer Hydraulics.

cylinder, for example—will create a vortex or whirlpool in the reservoir. If there is insufficient oil, air will be sucked into the system that will cause the pump to cavitate.

Extended use of equipment with oil at a low level also causes water condensation and the formation of sludge. Both cause the corrosion of hydraulic components and greatly accelerate wear.

Low oil level also speeds wear by increasing the concentration of dirt in the system. High pressure and/or high temperature operation increases any oil leakage. Oil leaking from the system is relatively clean because of the straining or filtering effect of sealed clearances. Leaks not only reduce oil level but concentrate the dirt in the hydraulic system.

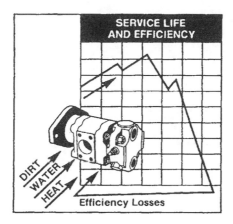

Figure 19.6 Dirt, water, and heat are the key factors in efficiency loss. Courtesy of Commercial Intertech.

The cure for low oil is not simply adding new oil. You must find and correct the causes. Educate your operators to watch for leaks at fittings, seals, and packings, and to report any leaks right away.

Filters, Fillers, and Breathers

A good filter helps by removing dirt from the hydraulic system, but you cannot depend on a filter alone to ensure clean oil. Most systems use a by-pass arrangement that permits oil flow even if the filter becomes clogged. Any oil that goes through the by-pass carries dirt along with it, causing accelerated wear on all system components. Change oil and filters on a regular maintenance schedule. It pays.

Do not overlook the danger of dirt accumulating around the oil filler or the unhappy results of a loose or lost filler cap or breather. Both allow dirt to enter the system. Make sure filler caps and breathers are clean and securely in place.

And, do not forget to keep the oil storage tanks, transfer pumps, and containers clean. Contamination can enter the system in many ways. The result of dirt is wear and the more dirt, the more wear.

Heat Build-Up

Heat, like dirt, compounds wear in a hydraulic system. Heating thins the oil which increases friction, wear, and leakage. It also breaks down oil causing sludge and varnish to form. Worn parts reduce efficiency and causes additional heat. Heat build-up can be reduced by keeping the oil clean and at the proper level and by replacing worn components which reduce efficiency.

Corrosion is another prime cause of wear and it is accelerated by heating. Each 10-degree rise in oil temperature actually doubles the rate of corrosion on exposed surfaces.

Make wear control an integral part of your regular preventative maintenance program. In summary—dirt, water, and heat are the key factors in efficiency losses (Figure 19.6). An efficient preventive maintenance program is cheap insurance.

20
Troubleshooting Exercise:
Bar Trimming Machine

Problems in hydraulic systems generally can be defined as relating to flow, pressure, or direction of flow. They manifest themselves in the form of noise, vibration, heat, or leaks. These manifestations can frequently be detected by the human senses of sight, touch, and hearing. These senses are controlled by our brain, which has a great capacity for analyzing and arriving at judgment decisions. Thinking is the starting point for troubleshooting. Acting follows immediately thereafter when the thought process has reached the mature judgment stage.

To establish a starting point for searching out the trouble requires knowledge of several different orientations. Troubleshooting requires knowing the functional design of the hydraulic system and what it must do correctly to assure that the machine will perform to its design specifications. This includes the force required from each actuator, the operating sequence of each motion, and the cycle time elements for a complete production cycle.

Problems with the system relate directly with one or more malfunctioning components. The objective is to quickly and accurately determine which component or series of components is malfunctioning. There is a logical sequence of thought and action, all of which requires an intimate knowledge of:

- hydraulic terminology (introduced in Chapter 1 and continued throughout the component study)
- component graphic symbols (energy flow, line technology, and component symbology introduced in Chapter 1 and continued in later chapters with the in-depth study of each individual component and its symbol within the hydraulic system)
- circuit diagram interpretation know-how (gained from hands-on experience after study of components and after watching the machine in normal operation)
- machine operational performance (learned from watching machine in normal operation and/or interrogation of operators and production supervisors)
- hydraulic component functions (explained in detail in previous chapters)
- troubleshooting experience (the summation of component knowledge and hands on activities)

Add to this knowledge the mental discipline to think first before acting. In addition to determining what is at fault, consideration must also be given to why the problem occurred and what should be done to keep it from happening prematurely again.

A CIRCUIT DRAWING SAVES TIME AND EFFORT

A good starting point in the troubleshooting process is to obtain the circuit diagram. Reviewing the circuit provides the opportunity to visualize how the individual hydraulic components relate to both machine performance and the interrelationship of one component to another. It is much easier and more rewarding to have a general overview of the hydraulic system by looking at a drawing.

Attempting to visualize the hydraulic system of any but the simplest machine and how it functions by looking at the machine first can be self-defeating. As many of the major hydraulic components are not visible from a single vantage point, time can be lost simply in becoming familiar with the hydraulic system. This time is better spent reviewing the circuit while relating the trouble symptoms to which components could be at fault. Another factor is that the troubleshooter is under pressure to arrive at a solution. Self-confidence gained by keeping a clear mind and approaching a solution in a logical fashion contributes to an early and successful conclusion.

Unfortunately, in many manufacturing facilities, it is unusual for the circuit diagrams to be readily available at the machine. More than likely they are stored at another place and are difficult, if not impossible, to find. The easy availability of clear plastic laminates and envelopes makes it possible to keep drawings on or near the machine. Managers of production facilities where critical manufacturing machinery must be kept running or face very expensive downtime have opted to make the availability of circuit drawings at the machine compulsory.

THE PITFALLS OF ACTION WITHOUT THOUGHT

To demonstrate the pitfalls of "act first" then "think," it is quite common to blame the pump first. Something has gone wrong and the pump seems to be a logical selection for investigation. Common sense seems to say that if an actuator slows down or stops, then it must be the pump. If the troubleshooter bends to the temptation of replacing the pump with no additional thought applied, then we are into the traditional "hit or miss" mode of troubleshooting. This mode in the majority of circumstances leads to a long and costly process of locating the fault and solving the problem. There are better ways, by far, than the "hit or miss" approach. One such way is described in a troubleshooting exercise which follows.

THE BAR TRIMMING MACHINE

The bar trimming machine (Figure 20.1) consists of a carriage that is driven in two directions by a hydraulic motor. Mounted on the carriage is a traverse cylinder that moves a cutting torch across a steel bar. Two clamp cylinders hold the bar in place and a semirotary actuator movers the cut off bar onto a roller conveyor at the end of each cut-off cycle.

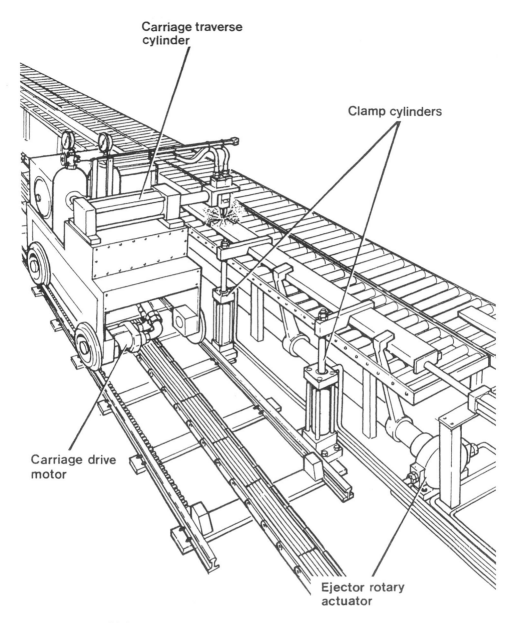

Figure 20.1   The bar trimming machine. Courtesy of Vickers, Incorporated.

The operator has reported three problems that occurred simultaneously. Prior to this, the machine had been working satisfactory for a long time.

Symptoms:

1. The carriage drive has slowed down in both directions.
2. The cutting torch traverse cylinder is operating slowly when extending.
3. The system is running hotter than usual.

The logical starting point for the troubleshooter if the circuit diagram (Figure 20.2) is not on or near the machine is to:

1. Obtain the hydraulic system circuit diagram (Figure 20.2) and become knowledgeable with the components and how they must function in order to assure correct machine performance. If the hydraulic circuit diagram is at a protected location on or near the machine considerable time and energy can be saved.
2. Review the problem with the machine operator to become familiar with its performance during normal operation and to discuss the problems in more detail.

Note on the circuit diagram that each hydraulic component is identified by a number inside of a circle. We will call these the I.D. (identification) numbers. Many circuit drawings will have these I.D. numbers related to a bill of materials. In the review and study process keep the circuit drawing clean and legible. A piece of clear plastic over the drawing may be a prudent investment.

The symptoms described by the machine operator provide the clues for the troubleshooting process. Two different actuator faults have occurred, both of which relate to "flow." The carriage drive has slowed down in both directions and the cutting torch traverse cylinder is operating slow when extending.

A common denominator to both problems could relate to the operator sensing that the hydraulic system is running hotter than usual. Since two faults have occurred simultaneously, the question arises as to whether or not a single fault is related to both problems. If the pump was at fault, this could bear on the slowing down of the carriage and the traverse cylinder. However, the operator did not report any degradation of the ejector or the return of the traverse cylinder. The clue of increased system oil temperature is of major significance.

The facts are that there are several components in the hydraulic circuit related to the three clues. For the experienced troubleshooter, the clues supplied point to a definitive action plan. However, for the less experienced troubleshooter, prudence suggests a detailed analysis be made to assist in thinking the problem through. With time and experience, shortcuts to the solution will develop.

Starting Point:

Begin the analysis by coding each hydraulic component on the circuit diagram such has been done already on the circuit attached. Focus on each component that is malfunctioning separately.

Machine Fault #1: Slow speed of the carriage drive.

Step 1. The symptom is one of speed; therefore, the problem is flow to the drive motor.
Step 2. Review the circuit diagram and establish the function of each component. Understanding the operational function of each component and their relation to the machine performance is key to troubleshooting.

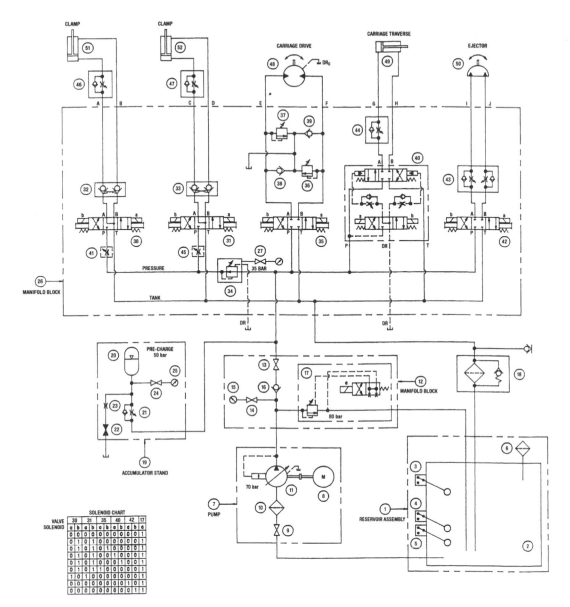

Figure 20.2 Bar trimming machine hydraulic circuit.

Step 3.   List by their code number each hydraulic component that could affect the carriage drive motor speed.

The objective is to narrow down the number of hydraulic components that could contribute to the problem. During the learning process, do not attempt to apply too much judgment, as it is better to have too long a list than too short a one, since the latter could leave out the problem unit.

Referring to the circuit diagram, the hydraulic units listed below could cause the problem. The comments related to each component provide an insight as to the thought process involved in their selection.

Component number	Comments
9.	A partially closed inlet valve could starve the pump and hence the system of fluid (but under these circumstances the pump would most likely be cavitating and thus noisy).
10.	A blocked inlet filter could cause the same effect (low output from the pump will affect the speed of the carriage drive motor).
13.	A partially closed shut-off valve will have the same effect.
16.	A partially open check valve (stuck due to contamination) will also have the same effect.
17.	A leaking (internally) relief valve will also have the same effect.
20.	It can be assumed that the purpose of the accumulator is to supplement the pump flow. An incorrect precharge pressure can affect the flow rate from the accumulator.
21.	An incorrectly set flow control valve would restrict the flow from the accumulator and thus to the motor.
22.	A blow-down valve left partially or fully open will reduce the flow both from the accumulator and the pump.
30 & 31.	A directional valve leaking internally from "P" to "T" can reduce flow to the motor.
34.	An excessive leakage down the drain line of the pressure reducing valve can reduce flow to the motor.
35.	A restriction in the directional valve controlling the carriage drive motor (caused by contamination or scoring), resulting in incomplete spool movement, can reduce the flow to the motor.
36 & 37.	Although a leaking relief valve would affect the flow to the motor, two faults would have to occur to affect the speed in both directions (i.e., either 36 and 37 leaking, or 38 and 38 leaking, or 37 and 39 leaking); while this is not impossible, it is probably quite unlikely.
40.	Same as 30 & 31 above.
42.	A worn or damaged motor having excessive case drain leakage will reduce effective flow through the motor.
49 & 50.	From the solenoid energization chart, neither the traverse cylinder nor the eject rotary actuator are operating at the same time as the carriage drive motor; therefore, an internal leak in either of these two actuators will not affect flow to the motor.
51.	Because the clamp cylinders are held down under pressure when the carriage drive motor is operating (refer again to solenoid energization chart) an internal leak across the piston seals can affect the carriage drive motor.
52.	Same as in 51 above.

Having completed the list of possible contributors to the slow down of the carriage drive motor, prepare a similar list of components relating to the traverse cylinder. Do not be concerned about duplicating components already being considered for the carriage drive motor problem.

Machine Fault #2: Slow speed of the traverse cylinder when extending.

Step 1.   The symptom is one of speed; therefore, the problem is flow.
Step 2.   As in machine fault #1, review the circuit diagram and establish the function of each component.
Step 3.   List the units that could affect the flow to the traverse cylinder when extending.

As in machine fault #1, the comments below relate to the components selected for consideration. In the list below, many items are the same as for machine fault #1 and there are additional components to be evaluated.

Unit number	Comments
9.	See Machine Fault #1.
10.	See Machine Fault #1.
11.	See Machine Fault #1.
13.	See Machine Fault #1.
16.	See Machine Fault #1.
17.	See Machine Fault #1.
20.	See Machine Fault #1.
21.	See Machine Fault #1.
22.	See Machine Fault #1.
30.	See Machine Fault #1.
31.	See Machine Fault #1.
34.	See Machine Fault #1.
35.	See Machine Fault #1.
40.	See Machine Fault #1.
42.	See Machine Fault #1.
44.	Failure of the bypass check valve (i.e., jammed partially closed) in the flow control valve will restrict the flow to the traverse cylinder when extending.
49.	A leak across the piston seals of the traverse cylinder will reduce the effective flow rate to the cylinder.
51.	See Machine Fault #1.
52.	See Machine Fault #1.

Having identified the components relating to each fault, plus the fact that both problems relate to flow, it becomes obvious that there are certain components that could contribute to the problems.

There are 52 components in the system, with 17 common to both: 9, 10, 11, 13, 16, 17, 20, 21, 22, 30, 31, 34, 35, 40, 42, 51, and 52. Narrowing down the number of components from 52 to the 17 most likely candidates highlights the importance of being able to understand the circuit and to put it to use in the thought process leading to a solution. Knowing that there are components common to both faults helps define the direction of investigation.

Step 4. Arrange the list of components in order of checking. Although the order is arbitrary, it may be influenced by such things as past experience, layout of components, position of gauges, etc.; however, some units will be easier to check than others.

Step 5. Preliminary check. Before going to the trouble of fitting additional pressure gauges, flow meters, etc., or removing piping, there are certain things that can be checked with the instrumentation already in place or with the senses of sight, feel, and hearing; thus, these should be checked first.

1. Shut-off and flow control valves are in the correct position (components 9, 13, 21, 22).
2. A visual indicator (if present) of the suction inlet filter (component 10).
3. The pressure gauge in the pump outlet port line to check the setting of the pump compensator and relief valve (components 11, 17).
4. A pressure gauge in the reducing valve (component 34).
5. The directional valves can be checked for any abnormal symptoms (components 30, 31, 35, 40, 42).
6. The clamp cylinders can be checked for any abnormal symptoms (components 51, 52).
7. The check valve can be checked for abnormal symptoms (component 16).
8. Compile and order the complete list: components 9, 13, 21, 22, 10, 11, 17, 34, 30, 31, 35, 40, 42, 51, 52, 16.

Checking the clamp cylinder (52) reveals the problem—the cylinder is hot, especially at the rod-end port. The human sense of feel has led to the detection of the fault.

Step 5. Correcting the problem. Having discovered excessive heat at the rod end of the cylinder, it can not be proven that this is the problem. Remove the rod end connection, drain off the fluid, then pressurize the blank end. This should be done, in this case, while the cylinder is at the fully extended position. If the fluid continues to flow while pressure is on the blank end, this is evidence that the piston seal has failed. The cylinder can now be repaired or replaced.

Step 6. Think. Having repaired or replaced the unit, thought should be given to both the cause and the consequence of the failure. Assuming that the piston seal had failed, these are questions that should be addressed before the machine is restarted:

Was the cause of the damage to the seals contamination, wrong seals, wrong installation, poor seal design, or something else?
Will the failure have any effect on the rest of the system?

If the seal has broken-up, have seal particles entered the system?
Has the overheated oil become oxidized?
Have any valves or other components been adjusted to attempt to compensate for the leak and will now require readjusting?

By thinking about the cause and the consequence of the failure, the same or consequential ones may be prevented in the future.

SUMMARY

The bar trimming machine exercise was purposely simplistic. It did, however, contain the basic elements of "thought" then "action." The objective was to demonstrate that taking the time to become acquainted with the problem by reviewing the circuit diagram and discussing the problem with the operator were beneficial first steps. Next came an evaluation of what elements in the hydraulic system were logical candidates for investigation. This was followed by a methodical plan to focus on the problem rapidly and with a good degree of effectiveness. In addition, as is so often the case, the use of the human senses can make a significant contribution to the troubleshooting process.

21

The Utility Basket Truck

The origin of the ladder to climb to locations above the nominal ground level probably predates recorded history. The clumsy step ladder was a modification of the early peg and pole or rail and rung structures. The later invention of the extension ladder served the emerging utility industries plus fire and rescue operations. The ladder was a basic tool used with back breaking regularity for rescuing individuals trapped in burning buildings and from other hazardous locations as well as the harvesting of wild honey and the construction of dwellings. A virtual river of paint has been applied with the aid of the ladder to reach elevated areas. Cats rescued from trees via the extension ladder are legends in small town America.

THE HYDRAULIC ASSIST

Fluid power systems using petroleum oil as the power transmitting media for military purposes in the very early years of the 20th century established the credibility of this discipline. Naval applications in particular provided the basis for the use of hydraulic and pneumatic power for force and motion requirements in the operation of construction machinery. This soon branched out into railway maintenance machinery and a host of other applications where portability, versatility, and dependability were of primary concern.

MOVING PEOPLE TO THE WORK PLACE

Thus there emerged force and motion mechanisms to position personnel. The special ladders, buckets, and platforms for the many different work functions involved in personnel movement are primarily positioned by means of actuators powered and controlled with hydraulic pumps and control valves.

Fluid power used with agricultural machinery found its basic and first use for controlling the position of the plow for ground preparation. Within a few decades the agricultural applications spread in many different areas. As an example, the platform vehicle of Figure 21.1 provides an ideal mechanism to position the operator for pruning,

Figure 21.1 Mobile man lift for orchard servicing. Courtesy of Prune-Rite, Modesto, CA.

spraying, harvesting, and inspecting the crop for decay or to establish the harvesting schedule. The relatively small power unit is easily accommodated within the basic framework of the machine and all of the control valving is contained within a metal block with cartridge valving for some functions and directional valves manifolded to the basic host valve block (Figure 21.2).

THE VERSATILE POWERED LADDER

Fire fighting professionals have two basic tasks to perform. One is to evacuate personnel from hazardous locations and the second is to position the stream of water advantageously to contain the conflagration. Various configurations of a basic moveable and extendable ladder and/or bucket are available to provide the needed service. Hydraulically actuated outriggers provide stability for the ladder structure. The photograph of Figure 21.3 shows several powered ladder and basket trucks in action.

The engineering, service, and maintenance drawing of the hydraulic circuit for the 95-foot "Stratospear" platform vehicle is divided into two basic sections. The first relates to the retractable outrigger structure, possible winch and hose storage devices, and the hydraulic pump and reservoir on the machine. The second section relates to those items located on the rotatable table to which the ladder assembly is affixed. The circuit drawing for the rotatable assembly provides a bill of material that includes the quantity of each component used, identification of the item, and a corporate part number assigned by the

Figure 21.2 Valve assembly for the machine illustrated in Figure 21.1. Courtesy of Hydro-Power Systems, Incorporated.

machine builder. Identification of the parts procured from vendors is also included in this list. Each item is also given an identification number that is shown in the parts list and adjacent to each component symbols on the drawing.

A typical parts list is shown in Figure 21.4. This drawing and parts list relates to the hydraulic circuit supplied through a swivel mechanism identified as item 25 or 26. A 0.75-in. pressure supply line and a 0.75-in. pressure return line passes through this swivel mechanism from the power unit on the truck chassis to supply the hydraulic circuitry above this swivel mechanism (Figure 21.5). All of the hydraulic circuit beyond this swivel is self-contained within the rotating ladder/bucket structure. Note the pressure gauge (10); pressure beyond the swivel can be read on this gauge which is protected by flow fuse (11) that will stop flow in the event of line breakage.

THE UNLOADING FUNCTION

The unloading valve in the lower right hand corner of Figure 21.6 connects the supply line to tank through the swivel at a pressure of approximately 100 psi. For emergency supply pressure in the event of a pilot malfunction the unloading valve can be negated by closing the manual override valve (15, Figure 21.5). Note the shuttle valve network shown in Figure 21.6. A separate, normally open, regulating valve is installed in each supply to the directional valve for the rotation, extension, and elevation functions. Note the pilot connection

Figure 21.3 A 95-ft boom extension "Stratospear" fire truck in action. Courtesy of Emergency One, Incorporated.

between the outlet of the regulating valve and the directional valve. Pressure in excess of the bias spring at the outlet of the regulating valve will reduce the flow through the regulating valve by moving the spool towards the bias spring. The pilot connection to the shuttle valve that senses the highest load in the cylinder lines is connected to the spring cavity of the regulating valve; this automatically senses the line with the highest working pressure. When the directional valve is in neutral there is no supply pressure to the shuttle valve. When the directional valve is shifted to a working position there is flow and pressure to the shuttle valve which is reflected to the spring cavity of the regulating valve. Thus the pressure drop across the directional valve will equal the pressure resulting from the bias spring in the regulating valve. The load that is sensed in the cylinder line is reflected as an additive to this bias spring in the regulating valve.

MULTIPLE SENSING LINES

The sensing line from the shuttle valve at the directional valve controlling rotation is also connected through another shuttle valve network at the directional control valve that

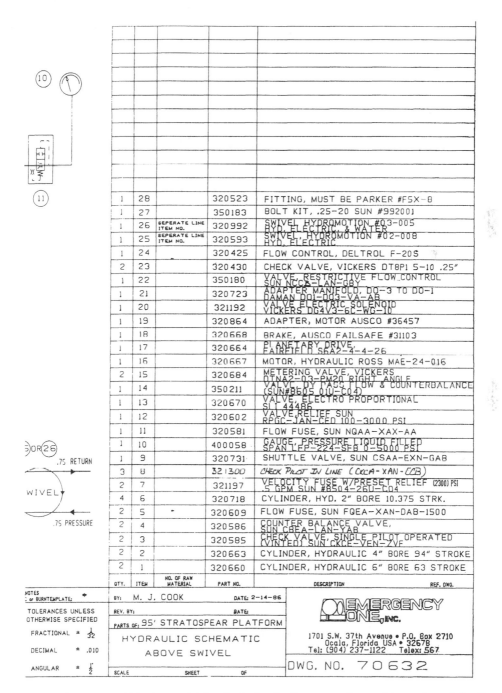

QTY.	ITEM	NO. OF RAW MATERIAL	PART NO.	DESCRIPTION	REF. DWG.
1	28		320523	FITTING, MUST BE PARKER #F5X-8	
1	27		350183	BOLT KIT, .25-20 SUN #992001	
1	26	SEPERATE LINE ITEM NO.	320992	SWIVEL HYDROMOTION #03-005 HYD, ELECTRIC, & WATER	
1	25	SEPERATE LINE ITEM NO.	320593	SWIVEL, HYDROMOTION #02-008 HYD, ELECTRIC	
1	24		320425	FLOW CONTROL, DELTROL F-20S	
2	23		320430	CHECK VALVE, VICKERS DT8P1 5-10 .25"	
1	22		350180	VALVE, RESTRICTIVE FLOW CONTROL SUN NCCA-LAN-GBV	
1	21		320723	ADAPTER MANIFOLD, DO-3 TO DO-1 DAMAN DO1-DO3-VA-AB	
1	20		321192	VALVE ELECTRIC SOLENOID VICKERS DG4V3-6C-WG-10	
1	19		320864	ADAPTER, MOTOR AUSCO #36457	
1	18		320668	BRAKE, AUSCO FAILSAFE #31103	
1	17		320664	PLANETARY DRIVE FAIRFIELD S6A2-4-4-26	
1	16		320667	MOTOR, HYDRAULIC ROSS MAE-24-016	
2	15		320684	METERING VALVE, VICKERS DTNA2-03-PM20 RIGHT ANGLE	
1	14		350211	VALVE BY PASS FLOW & COUNTERBALANCE (SUN#8605 01U-C04)	
1	13		320670	VALVE, ELECTRO PROPORTIONAL SI1 44486	
1	12		320602	VALVE,RELIEF SUN RPGC-JAN-CEO 100-3000 PSI	
1	11		320581	FLOW FUSE, SUN NQAA-XAX-AA	
1	10		400058	GAUGE, PRESSURE LIQUID FILLED SPAN LFP-224-SFB 0-5000 PSI	
1	9		320731	SHUTTLE VALVE, SUN CSAA-EXN-GAB	
3	8		321300	CHECK PILOT IN LINE (CKCA-XAN-CCB)	
2	7		321197	VELOCITY FUSE W/PRESET RELIEF (2300) PSI .5 GPM SUN #8504-26U-C04	
4	6		320718	CYLINDER, HYD. 2" BORE 10.375 STRK.	
2	5		320609	FLOW FUSE, SUN FQEA-XAN-DAB-1500	
2	4		320586	COUNTER BALANCE VALVE, SUN CBEA-LAN-YAB	
2	3		320585	CHECK VALVE, SINGLE PILOT OPERATED (VINTED) SUN CKCF-VEN-ZVF	
2	2		320663	CYLINDER, HYDRAULIC 4" BORE 94" STROKE	
2	1		320660	CYLINDER, HYDRAULIC 6" BORE 63 STROKE	

NOTES
: or BURNTEMPLATE:

BY: M. J. COOK   DATE: 2-14-86

REV. BY: DATE:

TOLERANCES UNLESS OTHERWISE SPECIFIED

PARTS OF: 95' STRATOSPEAR PLATFORM

HYDRAULIC SCHEMATIC

ABOVE SWIVEL

FRACTIONAL ± 1/32

DECIMAL ± .010

ANGULAR ± 1/2

SCALE   SHEET   OF

EMERGENCY ONE, INC.
1701 S.W. 37th Avenue • P.O. Box 2710
Ocala, Florida USA • 32678
Tel: (904) 237-1122   Telex: 567

DWG. NO. 70632

Figure 21.4 Typical bill of material for the hydraulically powered rotatable portion of the fire truck of Figure 21.3. Courtesy of Emergency One, Incorporated.

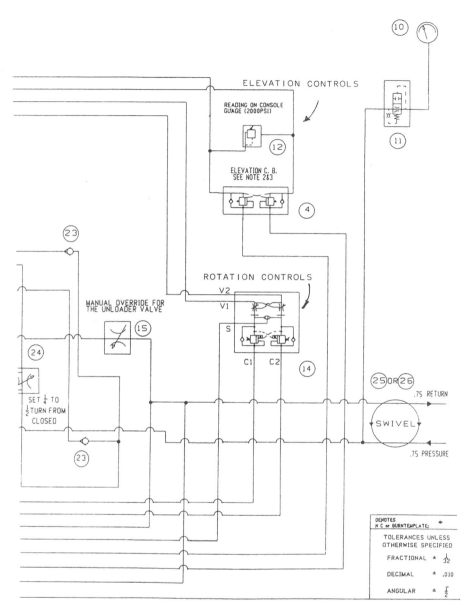

Figure 21.5 The swivel mechanism provides a dual rotary joint to direct pressurized fluid to the rotating devices and return spent fluid to the reservoir located on the truck chassis. Courtesy of Emergency Once, Incorporated.

controls extension. This network of shuttle valves employed to close the unloading valve does not impinge on the sensing function associated with the pressure drop across each of the directional valves involved with rotation, extension, and elevation functions. A connection from this unloading valve shuttle network passes through a check valve and flow regulating valve (24) in series (Figure 21.7). After passing through the flow regulating valve

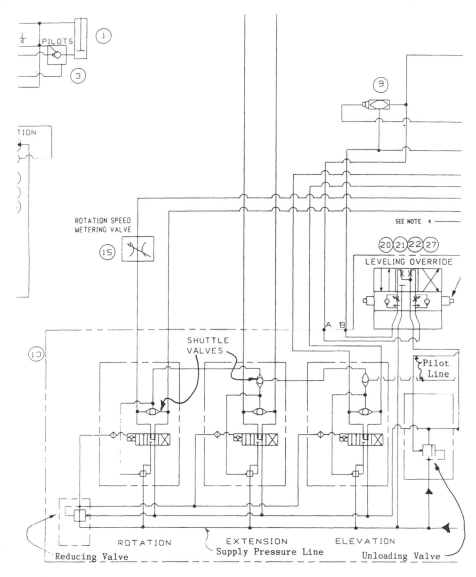

Figure 21.6 A reducing valve in the supply line to the major servo-type directional control pilot valves insures an appropriate pilot pressure level and uniform valve actuation. The unloading valve limits pressure and associated heat when movement is not needed. The shuttle valve network performs two functions. One is to sense the load at the actuator to limit pressure drop across the directional control valve. The second function is to provide a signal to close the unloading valve when force and motion is needed in any one of the circuits. Courtesy of Emergency One, Incorporated.

this pressurized oil is directed to the control chamber of the unloading valve adding to the bias spring value that moves the unloading valve to the closed position. The unloading valve will remain closed at all times when a pressure is sensed in any one of the actuator lines for the rotation, extension, elevation, or leveling function. A drawing note (Figure 21.5) states that restrictor valve 24 is to be set one-half to one-quarter from closed to provide

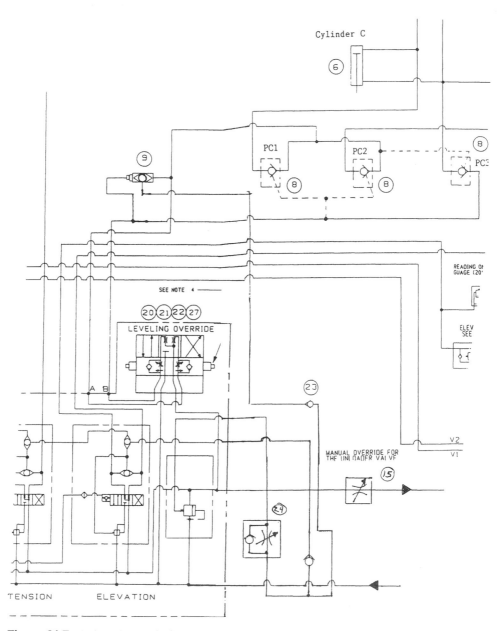

Figure 21.7 Valve 24 controls the rate at which the unloading valve closes to minimize shock as the specific hydraulic force and motion function begins. Courtesy of Emergency One, Incorporated.

the desired rate of loading the circuit. A shuttle valve (9) (Figure 21.7) accepts a pressure signal from either cylinder line to the leveling mechanism. This signal passes through a check valve (23) and a restrictor valve (24) to the control chamber of the unloading valve in a similar manner.

PILOT PRESSURE

Full line pressure is available to the pilot section of valve 20 controlling the leveling function via an internal connection. A reducing valve is installed in the common supply line to the servo-type directional control valves for rotation, extension, and elevation. A filter is also provided in each pilot line to the servo pilot section for these three valves. If a pilot line filter becomes saturated the piloted portion will not function properly. The electrical signal to the force motor actuating the valve should be checked thoroughly prior to replacing the servo valve. Servo valves are not normally considered to be repairable in the field. Because of their precision manufacture, servo valves normally require clean room facilities and trained technicians to make repairs and reassemble. Pressure beyond the reducing valve that supply regulated pressure to the servo valve pilot sections can also be checked with a gauge at the pilot pressure reducing valve. The pressure level is usually adjusted to a value between 100 and 200 psi. Control of the pilot pressure level insures repetitive smooth and accurate control of these force and motion functions.

FORCE AND MOTION

Four major force and motion functions are included in the rotatable segment of this fire truck. The hydraulic circuit clearly illustrates the flow of pressurized liquid to accomplish these force and motion functions. Leveling is accomplished by the use of Cylinders A, B, C, and D (Figure 21.8). A velocity fuse and integral relief valve installed in the head-end line of cylinders A and B protects the assembly from line loss in the event of a line break. The relief valve protects the assembly from excessive shock load. Rate of cylinder rod movement is controlled by pressure compensated adjustable orifices (22 on Figure 21.7) sandwiched in the manifold upon which directional control valve 20 in mounted. A note on the circuit drawing (Figure 21.7) instructs the service personnel to set speed metering for full stroke of leveling cylinders in *30 seconds*. A manual override is also integrated into the solenoid controlled, pilot operated directional control valve. Pilot operated check valve 1 (PC1) locks fluid in the head end of cylinders A and C (Figure 21.8). Pilot operated check 2 (PC2) locks fluid in the head end of cylinder B and D. A signal from the rod end is necessary to open pilot check 1 and 2. The fluid in the rod end of all four leveling cylinders is locked in by pilot check valve 3 (PC3). A signal from the head end is necessary to permit fluid flow from the rod end of cylinders A, B, C, and D. The physical location of the 2-in. bore leveling cylinders is shown in Figure 21.9. The relief and flow fuse assembly (7, Figure 21.8) protects cylinders A and B from excessive shock and line loss.

The Elevation Function

The head end of the elevation cylinders pivot at the base of the ladder assembly (1, Figure 21.9). Figure 21.10 shows the outrigger pads in position to stabilize and prevent tipping of

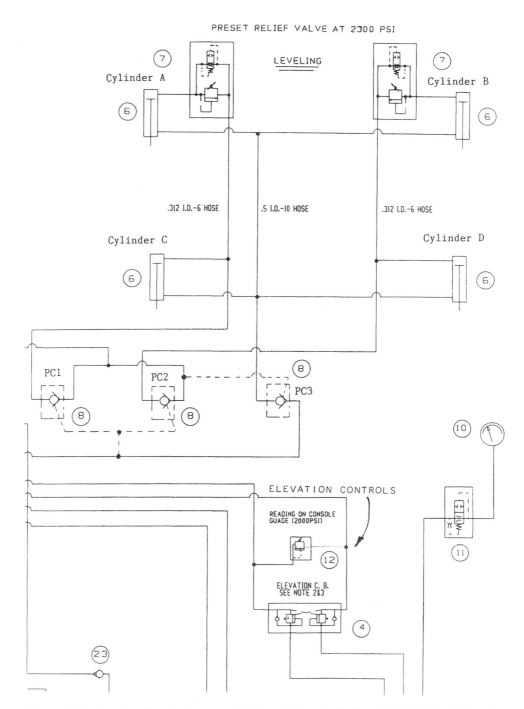

Figure 21.8 Leveling function is controlled by cylinders A, B, C, and D. Relief and flow fuse assembly protects the leveling cylinders from excessive shock and line loss. Courtesy of Emergency One, Incorporated.

Figure 21.9 Hydraulic motor, brake assembly, and leveling and elevation cylinders shown at the base of the rotating platform. Courtesy of Emergency One, Incorporated.

Figure 21.10 Outriggers in position ready to elevate the ladder assembly. Courtesy of Emergency One, Incorporated.

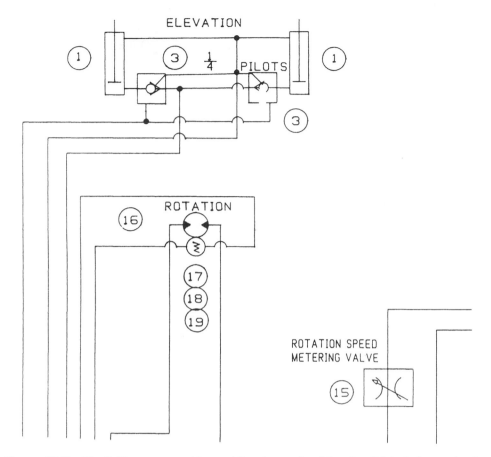

Figure 21.11   The fluid motor assembly providing the rotational function (16) includes a reduction gear assembly (17), a fail-safe brake (18), and an adaptor (18). The elevation cylinders are provided with pilot operated checks (3) that hold the cylinder in position until positive pressure is applied at the rod end of the elevation cylinders. Courtesy of Emergency One, Incorporated.

the ladder assembly. The two pilot operated checks (3, Figure 21.11) are inserted in the circuit at the head end of the elevation cylinders piped directly to the cylinder head (Figure 21.12). A relief valve (12, Figure 21.8) limits the pressure in the elevation function to a maximum value of 2000 psi. Counterbalance valves insure smooth movement of the elevation actuators in either direction of travel (4, Figure 21.8). A note on the circuit drawing (Figure 21.8) instructs the service personnel to adjust elevation counterbalance valve with no basket load at full retraction at 0–10 degrees elevation. The requested setting within the circuit drawing note is 1500 psi. Note the clevis connection at the rod end of the elevation cylinder assembly (Figure 21.12).

The Extension Function

Double-rod cylinders (2, Figure 21.13) extend and retract the ladder assembly. Smooth actuation and positive positioning is ensured by dual counterbalance valves (4). A note on

Figure 21.12 Note the pilot check valves connected directly to the head end of the elevation cylinders. Courtesy of Emergency One, Incorporated.

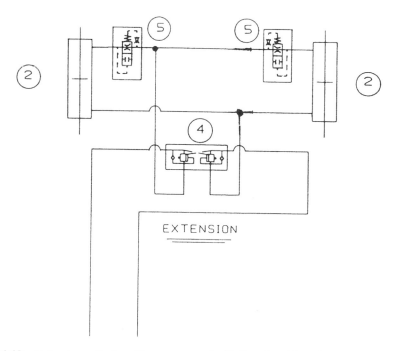

EXTENSION

Figure 21.13 Extension cylinders (2) are protected with flow fuse (5) and dual counterbalance (4). Courtesy of Emergency One, Incorporated.

Figure 21.14   Extension cylinders fastened to ladder assembly. Courtesy of Emergency One, Incorporated.

the circuit drawing instructs service personnel to set the counterbalance valve to 1000 psi. The position of the cylinders on the rotatable assembly is shown in the photograph of Figure 21.14.

The Rotation Function

Orbit motor (16, Figure 21.11) is coupled to a speed reducer and spring-biased position brake assembly. A pressure compensated bypass flow control and counterbalance valve is incorporated in valve assembly (14, Figure 21.5). A note on the circuit drawing instructs the service personnel to set the counterbalance valves at 1250 psi. The rotation speed metering valve is to be set for 180 degree rotation in 100-seconds at 1100 rpm. A shuttle valve is also included in valve assembly (14). The line from the shuttle valve to the brake release cylinder is identified by the letter S (Figure 21.5). The shuttle valve senses pressure which is directed to the brake assembly to release the spring actuated brake and allow the motor shaft to turn in response to the flow directed from the servo valve assembly. The counterbalance valves insure positive control of the motor function. No movement can occur until a pressure signal is received from the servo-type directional control valve. Pressurized oil directed into the line to the motor will first open the counterbalance valve and simultaneously cause the brake to release and the motor shaft to rotate at the rate established by the machine operator.

Control functions for the elevation, rotation, and extension are located in a console at

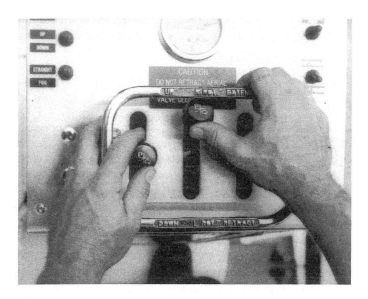

Figure 21.15   Control console for elevation, rotation, and extension functions. Courtesy of Emergency One, Incorporated.

Figure 21.16   Swinging the ladder assembly ready for stowing. Courtesy of Emergency One, Incorporated.

the base of the ladder/basket boom (Figure 21.15). Figure 21.16 shows a side view of the truck as the ladder assembly is swung toward the stowed position at the top of the truck cab.

DUAL BASKETS FOR SERVICING UTILITY LINES*

The Super/Stick (Figure 21.17) built by RO Corporation, Olathe, Kansas, is part of a new generation of equipment for handling overhead construction and maintenance for electric utilities. The 55-ft reach model has a total lift capacity of 3200 lb up to 30 ft away from the boom's centerline of rotation. Each of the twin baskets at the working end of the boom can rotate independently a full 180 degrees, and the operators in them have access to a lifting winch (Figure 21.18), separate conductor-lifting mast (Figure 21.19), and auxiliary hydraulic tools (Figure 21.20).

Pairs of cushioned cylinders, at the trunnion on the truck and at the knuckle where the two booms meet, move the booms to position the baskets. The natural geometry of this articulated machine (Figure 21.21), results in cyclical cylinder moments that generate a wide range of cylinder load pressures. During certain maneuvers, pressure can change quite rapidly. Because the upper boom, lower boom, and turntable functions may be operated simultaneously, the hydraulic system must be able to provide relatively constant flow to each function without regard to these pressure variations. A load-sensing hydraulic system meets this requirement to produce steady, smooth, and safe actuator speeds.

LOAD SENSING

The Super/Stick's load-sensing system operates on the principle that if a constant pressure drop is maintained across an orifice, flow through that orifice will remain constant. In this case the orifice is the passage through a directional control valve (as determined by valve spool position). The pressure drop is the difference between the inlet pressure to the valve and the load-induced outlet pressure at the valve's cylinder port.

The pump is an axial piston model (Figure 21.22), with both pressure- and flow-compensation. The pump's swashplate control spool is exposed to pump outlet pressure at one end and a combination of load pressure and a light spring at the other. Load pressure is delivered via a pilot line from the valve sense port to the spring end of the pump compensator. There, the load pressure is added to the spring pressure—in this case, approximately 400 psi.

Because the pressure forces on each end of the compensator must balance, the pump outlet pressure is adjusted to match this combination of load pressure and spring pressure. This assures that the pressure drop across the valve spool remains constant. Therefore, for any given spool displacement (or orifice) at the directional control valve, changes in load will not influence flow and actuator speed will remain constant.

---

*This material is reprinted from the March 1988 issue of *Hydraulics & Pneumatics* (March, 1988), Penton Publishing, Cleveland, Ohio. The author is R. J. Stallbaumer, Vice President, Engineering, at Williams Equipment Co., Inc., a distributor in Olathe, Kansas.

Figure 21.17 Twin hydraulic cylinders at Super/Stick's turret and knuckle provide outboard stability to handle heavy basket loads without rocking or swaying. Courtesy of RO Corporation, Olathe, KS.

Figure 21.18 An auxiliary winch can be provided at the work area adjacent to the baskets. Courtesy of RO Corporation, Olathe, KS.

Figure 21.19 Specialized auxiliary tooling can be accommodated in the basket work area. Courtesy of RO Corporation, Olathe, KS.

Figure 21.20 Heavy loads associated with the work requirements can be accommodated with the dual basket structure. Courtesy of RO Corporation, Olathe, KS.

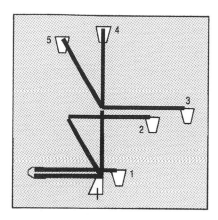

Figure 21.21 Some of the variety of possible upper and lower boom configurations that change load moments on the hydraulic cylinders. Courtesy of Williams Equipment Company.

DIRECTIONAL VALVES

The closed-center directional control valves at each operator's station are stack-type (Figure 21.23), with internal load-communication shuttles. Downstream load circuits are in parallel. The shuttle arrangement provides a pilot path that communicates the highest load pressure to the pump control, regardless of which valve spool is involved with that load. The pump delivers a pressure approximately 400 psi above the highest load pressure. The only power loss is the product of flow and compensating pressure for the highest pressure valve spool, plus the product of flow and difference between pump pressure and load

Figure 21.22 Variable-displacement axial piston pump response to pressure signals from its outlet and the load to maintain constant pressure drop across the directional control valve. Courtesy of Rexroth Corporation, Mobile Controls Division.

Figure 21.23 Stacked, noncompensated, directional control valves with integral shuttle logic have excellent metering characteristics. The system assures constant flow through the valve section handling highest load. Courtesy of Rexroth Corporation, Mobile Controls Division.

pressure for each activated lower-pressure valve spool. This, of course, minimizes heat build-up in the hydraulic system.

ADJUSTABLE LOAD CHECKS

Adjustable load checks at the valves' cylinder ports perform two functions. They establish the maximum flow (actuator speed) through each valve by their orifice area and they support the loads until pressure through the valve is high enough to move them.

Providing that there is sufficient flow to supply all spool requirements simultaneously, multiple-function operation is possible. Because the boom rotation turntable normally functions at relatively low pressure, the counterbalance valves on the rotation motor are set at approximately half the maximum boom cylinder pressure to provide optimum control during multiple-function operation.

LOGIC SELECTS ACTIVE CONTROL STATION

Each basket includes a complete set of controls (Figure 21.24); these are duplicated at the turntable turret below on the truck. These controls operate the boom movements, lifting winch, conductor-lifting mast, hydraulic tools, truck engine start/stop, and rotation of either basket. Obviously, only one set of controls can be in command at any time for safety reasons. A simple hydraulic logic circuit uses the pump's standby pressure (which is always available, even at idling speed) and NC *pilot-to-open switching elements*. The elements block or open flow paths in the high-pressure power circuit. The pilot logic circuit uses small valves and hoses, thereby reducing weight and cost.

Detented two-position, four-way valve A (Figure 21.25) lets the ground-level operator select a station or pass control to the baskets above. The operators in the baskets then can

Figure 21.24 Control panels are easily accessible to the operator. Courtesy of RO Corporation, Olathe, KS.

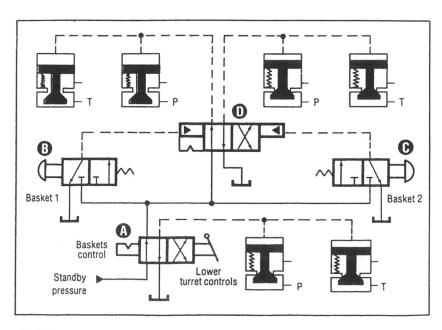

Figure 21.25 Logic circuitry uses normally-closed pilot-to-open switching elements. Courtesy of Williams Equipment Company.

use momentary signals at push button valves B and C to shift detented double-pilot valve D to activate one basket's controls and disable the other. With this arrangement, the operator on the ground can always take control if necessary, should the operators in the basket find themselves in trouble.

SUMMARY

To troubleshoot these machines it is necessary to understand that each force and motion function is a separate and distinct subcircuit within the total machine circuit. Failure of one force and motion function can be isolated from the others if it is the only one that will not function. If all functions become inoperative the basic pressurized fluid supply chain must be investigated. Adequate pressure at the pressure gauge (10, Figure 21.5) indicates proper pump supply function to this rotating portion of the machine. The next point of investigation would be the pilot pressure supply to the servo valves. The reducing valve in the supply line could be at fault. If the leveling function performs in a satisfactory manner the pilot supply to the servo valves could be creating the malfunction.

Malfunction of any component within a specific force and motion functional area can be resolved employing the troubleshooting techniques learned in previous chapters.

22

Backhoe and Loader

Flexibility of the many force and motion functions is an essential ingredient to accomplish the tasks performed by a tractor equipped with a backhoe and front end loader. Many different configurations are available for specific end use tasks that dictate machine size and capabilities. With the tools in a stowed position the machine moves quickly and easily from job to job (Figure 22.1).

The basic machine may be modified with a minimum number of changes to the physical structure for plowing or pulling ground preparation devices for agricultural purposes. It can serve as a front end loader to move bulk materials in a needed pattern.

Digging in a fixed area with hydraulically powered outriggers holding the basic machine in the desired location with appropriate stability may be the biggest challenge to the hydraulic circuit. Rapid bucket movement with continually changing force requirements are easily accommodated by the sophisticated hydraulic force and motion power transmission system.

Troubleshooting techniques are acquired by using the knowledge gained in the first part of this book. Necessary skills can be developed for communication purposes. A working knowledge of standard fluid power symbols is essential to read and understand the circuit drawings. The basic functions of pumps, control valves, and actuators have been explained. The additional information provided in this chapter outlines the specific characteristics of the components and how they relate to the force and motion functions of this versatile family of machines. Step-by-step procedures then provide the guide for pinpointing a specific malfunction and how to make the needed repairs.

COMPONENTS

Pressurizing the Fluid

Two gear-type pumps driven by a common shaft are employed to selectively pressurize the fluid to supply the force and motion circuits associated with power steering, loading, digging, three-point hitch, or other auxiliary functions (Figure 22.2). The rear section adjacent to the input drive shaft of the machine serves a specific purpose. This section of the

Figure 22.1   The backhoe/front end loader can travel from job to job with the working tools in the stowed position. Courtesy of Case Corporation.

pump delivers 24 U.S. gpm at 2000 psi at 2000 rpm directly to the loader and steering functions. A power-beyond connection directs pump delivery to the backhoe control valve if the loader or steering circuit is not calling for pressurized fluid. The front pump section delivers 9 U.S. gpm at 2000 psi at 2000 rpm to the backhoe or three-point hitch circuit (Figure 22.3).

A priority valve is located at the inlet section of the loader control valve. The priority valve is normally spring biased to direct pressurized fluid to the power steering control assembly. The flow from the large volume pump is available to the loader control circuit at all times. No fluid is used for steering when the power steering control valve is in a neutral or rest position. An appropriate flow is immediately available for power steering at the operator's demand. Only that portion needed for the steering function is diverted in the priority function and the remaining volume from the large pump is available to the loader control circuit.

The priority valve illustrated in Figure 22.4 shows the valve in the rest position (dead engine). When the machine is started the flow entering the inlet valve will first provide oil simultaneously to the power steering supply port and to the chamber between the spool (4) and the plug (5). If steering is not desired the pressure in the chamber will raise slightly until the spool is displaced against the spring (2) and the flow from the pump is directed to the loader. When steering is required a signal pressure enters the priority valve in the spring area and forces the valve into the position shown, diverting oil to the steering supply port until the demand is satisfied.

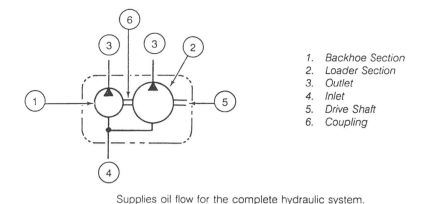

1. Backhoe Section
2. Loader Section
3. Outlet
4. Inlet
5. Drive Shaft
6. Coupling

Supplies oil flow for the complete hydraulic system.

Figure 22.2 Schematic and sectional view of pump assembly. Courtesy of Case Corporation.

The cartridge-type relief valve to limit the maximum pressure in the steering system is located at the inlet section of the loader control valve.

Power Steering

The upper diagram of Figure 22.5 illustrates by symbols the flow through the steering control circuit. The steering cylinder is a double-rod, push-pull assembly, with equal displacement in each direction. Note that valve (5) is actuated from the steering wheel. The symbol also shows that there is a mechanical feedback (6) from the metering device within the valve assembly. This metering device is shown in the lower view of Figure 22.5 in cross section. Flow to the steering cylinder is metered by this control valve assembly which

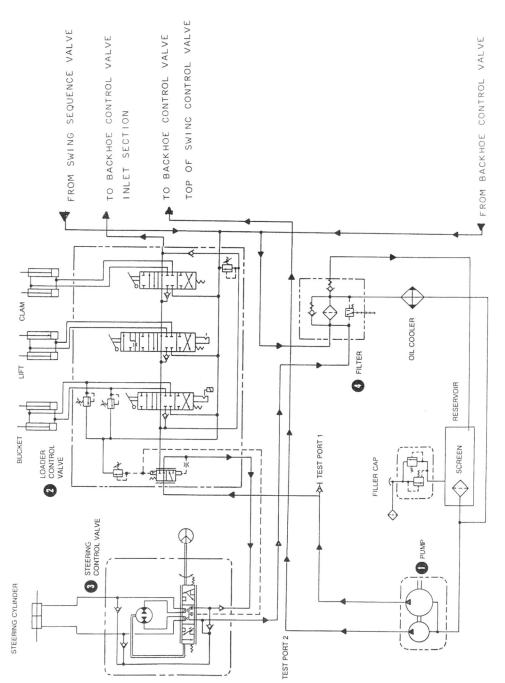

Figure 22.3 Pump, steering, and loader control valve assembly, and fluid conditioning circuit. Courtesy of Case Corporation.

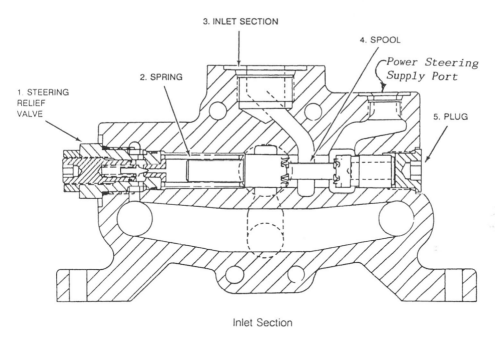

Inlet Section

Figure 22.4 A priority valve insures pressurized fluid supply for power steering system. A relief valve limits maximum pressure in power steering assembly. Courtesy of Case Corporation.

consists of a mechanical input from the steering wheel to a rotary servo valve as shown in section in the lower view of Figure 22.5. The flow to the metering assembly must pass through a generated rotor displacement set in series with the power flow to the steering cylinder. This maintains a finite relationship between the input movement of the steering wheel and the movement of the steering cylinder rod as it moves the steerable wheels. The gerotor-type (generated rotor gear set) displacement set (7) moves a sleeve within the housing assembly and cancels the input signal as the steering cylinder moves because of the controlled flow of oil through this metering structure (Figure 22.5). The power steering assembly can be replaced as a unit if foreign matter or mechanical damage creates need for a service function.

Loader Control Valve

The loader control valve is made up of five sections (Figure 22.6). Three separate and independent force and motion functions are controlled by this valve. Cylinders attached to the bucket provide needed movement pattern (Figure 22.7), lift cylinders can provide desired movement of the complete bucket assembly (powered or floating), and the clam cylinders (Figure 22.8) are integrated into the bucket assembly to selectively permit dropping the load in a desired area or into a truck. The inlet section of the loader control valve houses the power steering priority valve and power steering maximum pressure relief valve. The bucket section houses the directional spool, load checks, a relief valve for each cylinder line, and an electrical device (Figure 22.9). The electromagnet in the bucket valve is part of the return-to-dig feature in the loader circuit. After the bucket is dumped over the

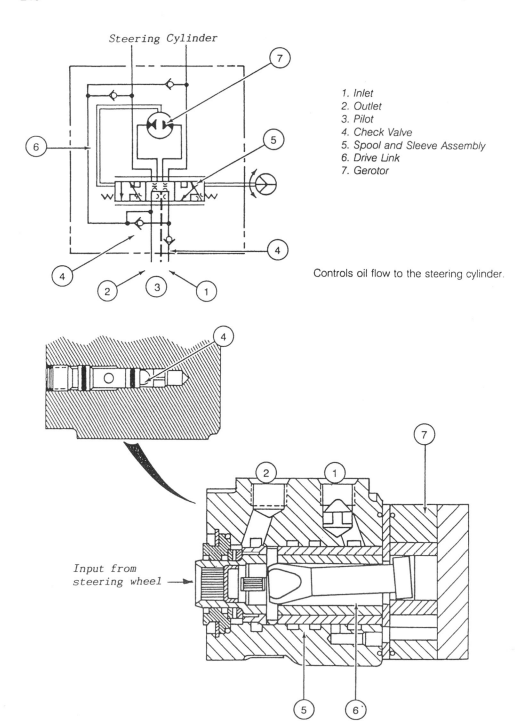

1. Inlet
2. Outlet
3. Pilot
4. Check Valve
5. Spool and Sleeve Assembly
6. Drive Link
7. Gerotor

Controls oil flow to the steering cylinder.

Figure 22.5 Steering control valve. Courtesy of Case Corporation.

Controls flow of oil to the cylinders for the loader by
movement of control levers.

1. Inlet
2. Outlet
3. To Steering Control Valve
4. From Steering Control Valve (Pilot)
5. To Bucket Cylinders
6. To Lift Cylinders
7. To Clam Cylinders
8. Main Relief Valve
9. Steering Relief Valve
10. Flow Control Valve
11. Circuit Relief Valve
12. Load Check Valve
13. Spool
14. Check Valve

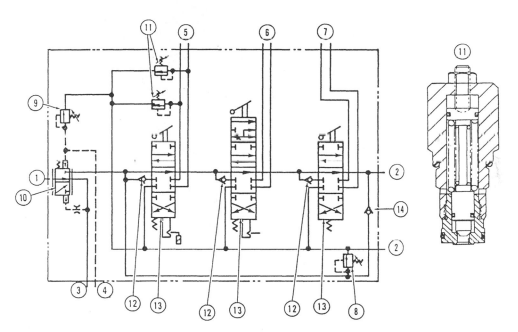

Figure 22.6 The loader control valve is made up of five sections including power steering, priority valve, bucket, lift, and clam cylinder directional control valves. The load checks are insert cartridges and the relief valves (sectional view, 11) are screw-in cartridges. Courtesy of Case Corporation.

discharge pile or into a truck the operator need only pull the control lever into the rollback position and the electromagnet will retain the spool until the bucket reaches the normal dig position (level to the ground). A switch is actuated at the predetermined stop to de-energize the magnet. This feature permits the operator to steer and perform other functions without concern for the bucket position. The switch is provided with an adjustable mounting feature to provide desired positioning of the bucket in the working mode.

Four positions are included in the lift directional control valve section. The functions include lift, hold, lower, and the fourth position provides a float condition. In the float

Figure 22.7 Relief valves are provided in the two lines to the bucket cylinder to absorb shock loads (trunion-mounted upper cylinders). The lift cylinders have the directional valve structured to permit a float position (lower cylinders). Courtesy of Case Corporation.

Figure 22.8 The clam cylinders are integrated into the bucket assembly. Courtesy of Case Corporation.

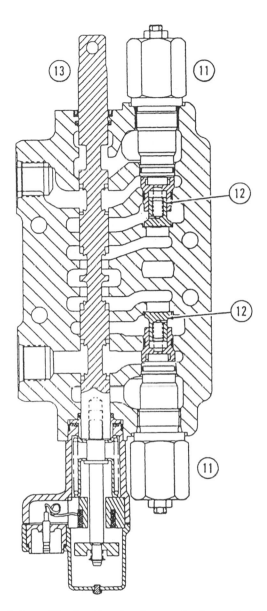

Figure 22.9 Sectional view of bucket directional control valve, cylinder line relief valves, load check, and electromagnet to provide electrical interlock. Courtesy of Case Corporation.

position the directional valve the two lift cylinder lines are connected to each other and to the tank through an orifice. Thus gravity and the terrain will determine the position of this working machine member. The clam section provides a conventional four-way directional control with load check valves to prevent cylinder movement when supply pressure drops because of other machine movements within the force and motion functions. The outlet section incorporates a check valve and the main relief valve (Figure 22.10).

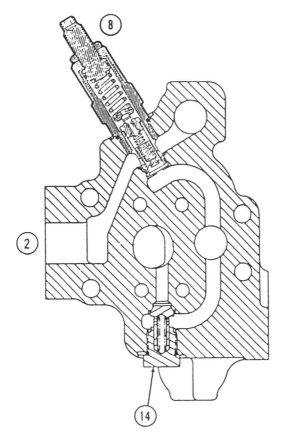

Figure 22.10 The cartridge-type main relief valve (8) for loader and backhoe functions is located in the outlet section of the loader valve. The check valve (14) permits high pressure oil in the power-beyond circuit (backhoe) to vent over the relief valve (8). Courtesy of Case Corporation.

Backhoe Control Valve

Seven directional control functions are incorporated in the backhoe control valve (Figure 22.11). The swing sequence valve is actuated by a cam structure; this controls the rate of movement in the force and motion pattern needed to swing the bucket assembly in a safe and efficient mode. Auxiliary valving within the backhoe control valve assembly is identified in Figure 22.12. A power-beyond connection from the loader control valve is connected by a suitable conduit to the inlet section of the backhoe control valve. Delivery from the smaller gear pump enters on the top of the swing directional control valve section and joins with the power-beyond from the loader control valve (Figure 22.13). Movement of the right and left swing cylinders is controlled by the first lever-operated directional control valve in the assembly. Meter-in restrictors are installed in the cylinder lines from the directional control valve to the swing sequence valve. Safety relief valves are installed in

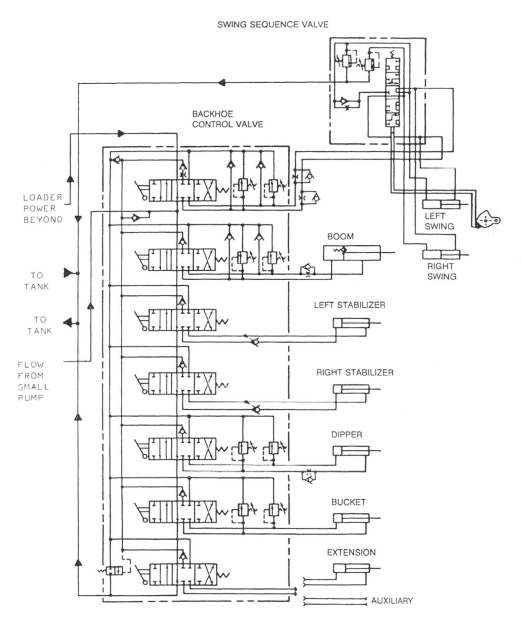

Figure 22.11   Backhoe control valve and swing sequence valve. Courtesy of Case Corporation.

the line between the head end of the swing cylinders and the swing sequence valve (Figure 22.14). Anticavitation check valves and maximum pressure relief valves are installed in the lines between the swing section directional control valve and the swing sequence valve. The relief and anticavitation checks are located in the body of the swing section directional control valve. Swing function cylinders are shown in the photograph of Figure 22.15 at the same level as the outrigger or stabilizer cylinders.

Controls flow of oil to cylinders for backhoe by movement of control levers.

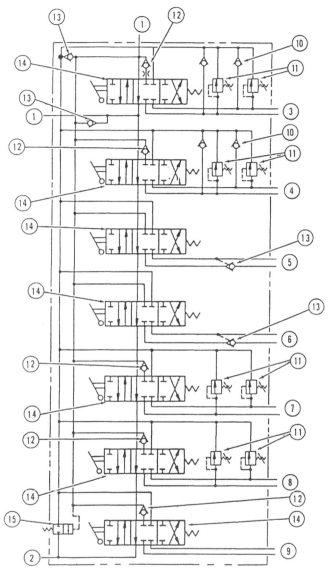

1.   *Inlet*
2.   *Outlet*
3.   *To Swing Cylinders*
4.   *To Boom Cylinder*
5.   *To Left Stabilizer Cylinder*
6.   *To Right Stabilizer Cylinder*
7.   *To Dipper Cylinder*
8.   *To Bucket Cylinder*
9.   *To Extension Cylinder*
10. *Anticavitation Valve*
11. *Circuit Relief Valve*
12. *Load Check Valve*
13. *Check Valve*
14. *Spool*
15. *Regeneration Valve*

Figure 22.12 Identification of auxiliary valving incorporated in backhoe control valve. Courtesy of Case Corporation.

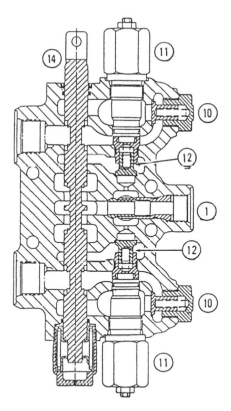

Figure 22.13 Swing section of backhoe control valve. Courtesy of Case Corporation.

Boom Section

Cylinder line relief valves and anticavitation checks are also included in the four-way directional control valve for boom control (Figure 22.16). A meter-out restrictor is inserted in the line from the rod end of the boom cylinder. A small relief valve is installed in the piston assembly of the boom cylinder. The relief valve serves to limit the pressure during end of cylinder stroke deceleration.

Left and Right Stabilizer Sections

A pilot operated check valve is integrated into the body of the left and right stabilizer sections to lock oil into the head end of the stabilizer cylinders (Figure 22.17). Flow to the rod end of the stabilizer cylinder to retract the cylinder provides pilot pressure to open the check valve in the head-end cylinder line.

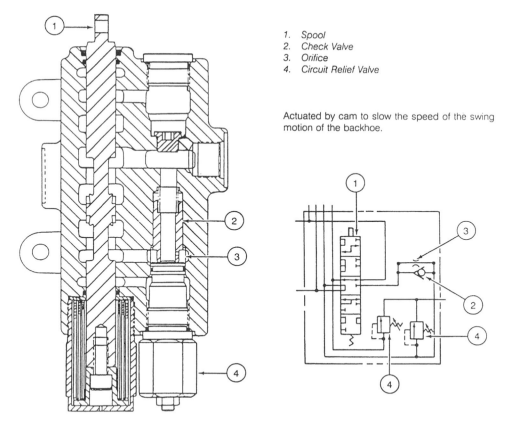

1. *Spool*
2. *Check Valve*
3. *Orifice*
4. *Circuit Relief Valve*

Actuated by cam to slow the speed of the swing motion of the backhoe.

Figure 22.14   Swing sequence valve. Courtesy of Case Corporation.

Dipper and Bucket Section

Maximum pressure relief valves are integrated into the cylinder lines within the valve body for both the dipper and bucket section directional control valves. Anticavitation checks are also included in the valve section body (Figure 22.18).

Extension Section and/or Auxiliary

Anticavitation checks are integrated into the directional control valve of the extension and/or auxiliary section (Figure 22.19).

Outlet Section for Backhoe Control

A pilot operated regenerative valve is integrated into the outlet section of the backhoe control valve (Figure 22.20). The regenerative valve is piloted to the open position by an operating pressure value sufficient to overcome the bias spring that pushes the valve to a closed position. Flow from a fast dropping cylinder can lower the pressure in the head end

Figure 22.15   Backhoe with stabilizer cylinders in position. Courtesy of Case Corporation.

sufficient to permit the closing of valve 15 (Figure 22.12). Return oil from the cylinder regenerates through check valve 13 into the major supply line adding to the supply from the pumps and avoids cavitation within this part of the circuit caused by the rapid movement of the cylinders (Figure 22.21).

Three-Point Hitch Control Valve

The three-point hitch control valve controls force and motion action for an auxiliary, tilt, pitch, and lift function (Figure 22.22a). The lift control valve is a four-position valve providing a float position. Supply to the three-point hitch control valve is provided by the small outboard gear pump (Figure 22.22b). The inlet section of the directional control valve assembly houses the maximum pressure relief valve (7, Figure 22.23). The directional control valve section for auxiliaries, tilt, and pitch include a load check (9, Figure 22.24).

The lift section directional control valve includes four working positions (Figure 22.25). In the neutral position cylinder ports are blocked and the supply passes through the valve to the reservoir. When shifted to the lift position a relief valve (8) is positioned in the valve body to establish a maximum pressure at the head end of the cylinder. An anticavitation check (9) is also included. When the lowering position pressure is directed to the rod end of the cylinder and that from the head end is diverted to the reservoir. A fourth position is provided to connect the two cylinder ports to each other and to the tank to permit the lift assembly to float with gravity with the terrain being the controlling agent. The position of the control valve, cylinders, and hitch assembly are shown in Figure 22.26.

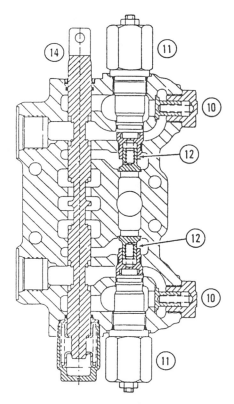

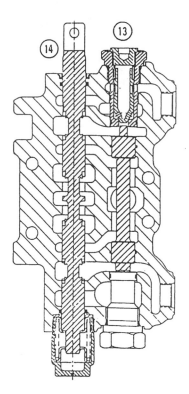

Figure 22.16 Boom section directional control valve. Courtesy of Case Corporation.

Figure 22.17 Left and right stabilizer sections directional control valve with integral pilot operated check valve to lock oil in the head end of the cylinder. Courtesy of Case Corporation.

Conditioning the Oil

The spin-on filter assembly of Figure 22.27 accepts all oil returned from the functional circuit for cleaning prior to being returned to the reservoir. Figure 22.27a illustrates the circuit components with ISO symbols. Bypass valve (6) provides an alternate path to the reservoir in the event of filter saturation or cold viscous oil that will not pass through the filter element in the normal pattern as the machine is started up in a cold environment. Oil flow passing through valve (6) will not be filtered and foreign matter in the oil will be returned to the reservoir. Pressure switch (7) provides an alert to the flow passing through bypass valve (6). Check valve (8) provides resistance to flow diverting a portion of the oil through the air-to-oil cooler prior to returning to tank. The numbered functions are listed in Figure 22.27b. Section view 22.27c shows the position of the bypass valve (6) and the resistance valve (8). Sectional view 22.27d shows the return from the steering valve assembly at (4) and the diversion port (3) connected to the cooler assembly. A sectional view of switch (7) is also shown in Figure 22.27d.

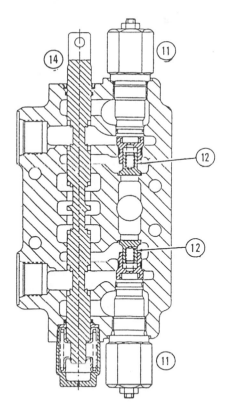

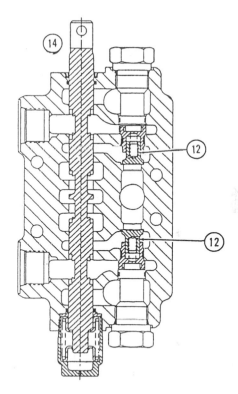

Figure 22.18 Dipper and bucket section directional control valve with integral anticavitation check valves. Circuit relief valves are installed in cartridge format. Courtesy of Case Corporation.

Figure 22.19 Extension section directional control valve with anticavitation check valves. Courtesy of Case Corporation.

The local dealer will usually recommend a filter change schedule appropriate to the operating conditions encountered when using the machine. The signal from the pressure switch will also alert the operator to need for service.

TROUBLESHOOTING FOR A PROBLEM IN ALL CIRCUITS

Having reviewed the circuitry for the backhoe and front end loader the flow chart of Figure 22.28 provides a guide for effectively seeking the component(s) that require service. Notes follow relative to the steps in the flow chart.

Step 1. Look for external leaks and indications of damaged links, hose, levers, cams, etc. Make repairs as required.

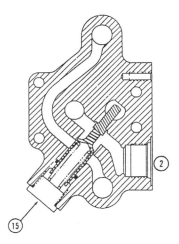

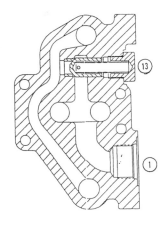

Figure 22.20 Outlet section of backhoe control valve with an integral regeneration valve. Courtesy of Case Corporation.

Figure 22.21 Check valve 13 (needed for the regenerative function) is located in the inlet section of the backhoe control valve. Courtesy of Case Corporation.

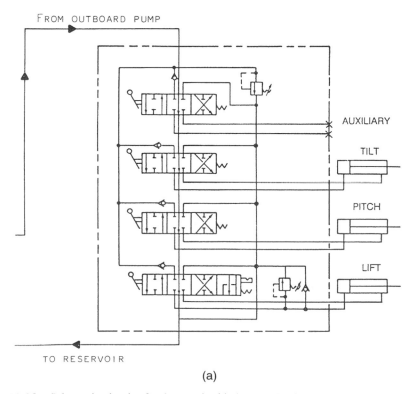

(a)

Figure 22.22 Schematic circuit of a three-point hitch control valve. (a) Control valve and cylinders. (b) Complete circuit. Courtesy of Case Corporation.

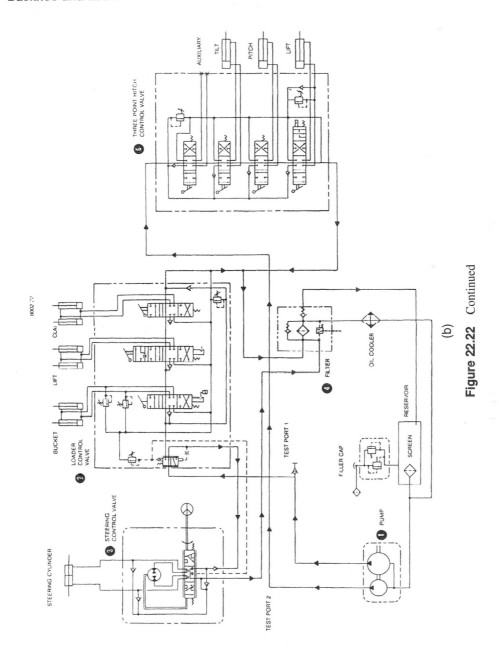

Figure 22.22 Continued

(b)

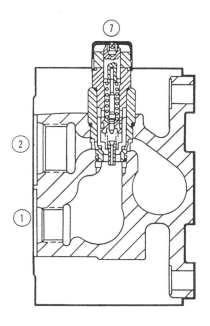

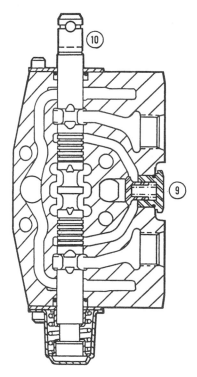

Figure 22.23 Sectional view of inlet section of a three-point hitch control valve with maximum pressure system relief valve. Courtesy of Case Corporation.

Figure 22.24 Sectional view of auxiliary, tilt, and pitch section showing integral load check. Courtesy of Case Corporation.

Step 2. When adding oil be certain that containers are clean and that the new oil has not been contaminated by loose or damaged caps during storage.

Step 3. Park the machine on a level surface. Put the backhoe in the transport position and lower the loader bucket to the floor. Heat the oil to operating temperature. Connect the 0–3000 psi (20,684 kPa, 207 bar) pressure gauge to the quick disconnect fitting (Figure 22.29). Run the engine at full throttle. Hold the loader control lever in the raise position until the loader stops moving. Continue to hold the loader control lever in the raise position and read the pressure gauge. Stop the engine. Compare the pressure gauge reading with the machine manufacturers specification. If the reading is not correct, adjust the main relief valve. Typical values are 2500–2600 psi (17,237–17,927 kPa, 172–179 bar). The main relief valve controls the pressure in the loader and backhoe circuit. It is located in the outlet section of the loader control valve. Loosen the lock nut on the main relief valve. Turn the adjusting screw clockwise to increase the pressure and counterclockwise to decrease the pressure.

Step 4. Replace the filter. The oil filter (5, Figure 22.27) cleans oil returning to the reservoir from the working portions of the hydraulic circuit. The indicator switch (7, Figure 22.27) signals need for a filter change. The suction screen (4, Figure 22.30) is designed to

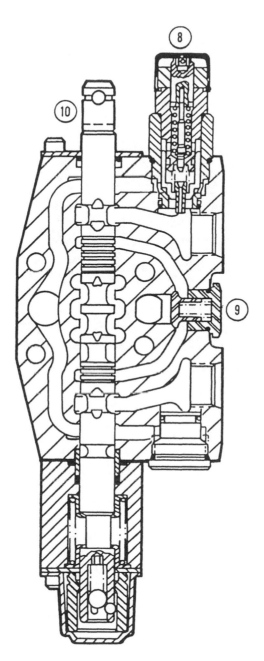

Figure 22.25 Sectional view of lift section showing maximum pressure relief valve and fourth spool position for float function. Courtesy of Case Corporation.

Figure 22.26 View of three-point hitch hydraulic components (i.e., control valve, cylinders, and control linkage). Courtesy of Case Corporation.

prevent ingestion of major contaminants into the pump inlet line that may have inadvertently entered the hydraulic reservoir.

Step 5. Do the stall test following manufacturers instructions. The stall test will indicate the condition of the torque converter, transmission, hydraulic system, and the engine. The stall test provides a method of finding the cause for poor performance. The engine is run at full throttle and the transmission and hydraulic systems are engaged separately, and then together. Comparing the engine speeds from the stall test with a check sheet supplied by the manufacturer will help to find the cause of the problem. It can be necessary to check a separate system to find the exact cause of the problem. The use of a phototachometer or other tachometer of equal accuracy is needed to get accurate results from the stall test.

The engine, transmission, and hydraulic system must be at operating temperature before doing the stall test. The oil must be heated according to instructions in the machine manual. Test results within the rated rpm range indicates that all systems are in good operating condition. Excessive or subnormal rpm indicates an engine problem. The rpm readings with various systems operating will selectively pinpoint trouble in the torque converter or hydraulic system.

Step 6. Check for contaminated oil. Contamination in the hydraulic system is a major cause of the malfunction of hydraulic components. Contamination is any foreign material in the hydraulic oil. Contamination can enter the hydraulic system in several ways:

1. when you drain the oil or disconnect any line
2. when you disassemble a component

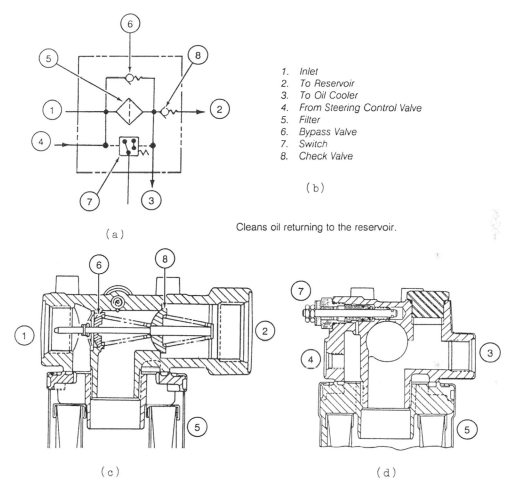

1. Inlet
2. To Reservoir
3. To Oil Cooler
4. From Steering Control Valve
5. Filter
6. Bypass Valve
7. Switch
8. Check Valve

(b)

Cleans oil returning to the reservoir.

(a)

(c)   (d)

Figure 22.27 Spin-on oil filter. (a) Symbol and parts identification. (b) List of component parts. (c) Sectional view from inlet to outlet. (d) Sectional view of line to steering control valve and outlet to cooler. Courtesy of Case Corporation.

3. from normal wear of the hydraulic components
4. from damaged or worn seals
5. from a damaged component in the hydraulic system

All hydraulic systems operate with some contamination. The design of the components in a typical backhoe system permits efficient operation with a small amount of contamination. An increase in this amount of contamination will cause problems in the hydraulic system. The following list includes some of these problems:

1. *Cylinder rod seals leak.* Inspect the rod for nicks or scores that may be damaging the packing. Polishing rough spots may extend the life of the packing and prevent contamination of the system. Replace seals as necessary. Use care in installing seals. Inspect for gland alignment and proper positioning of retaining devices.

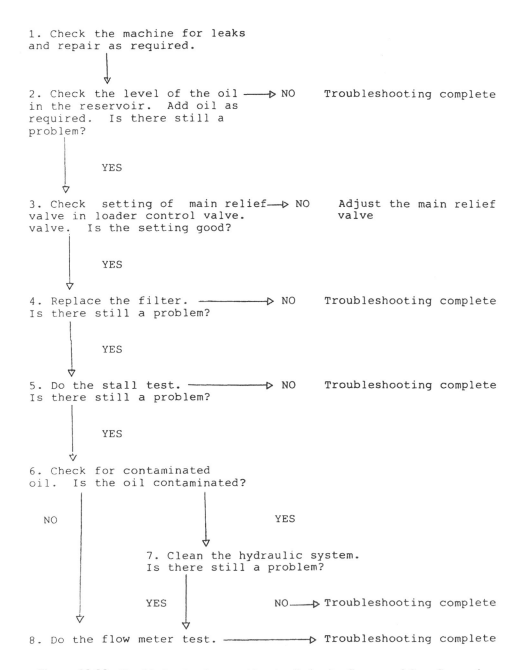

Figure 22.28 Troubleshooting for a problem in all circuits. Courtesy of Case Corporation.

2. *Control valve spools do not return to neutral.* Inspect for external damage to the spool linkage. Look for rust or jamming by foreign matter.

3. *Movement of control valve spools is difficult.* Inspect for damage to pivot pins and linkages. Inspect bore for scoring of body causing hydraulic unbalance. A replacement valve segment may be required.

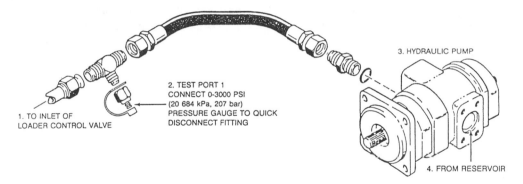

Figure 22.29 Connect the pressure gauge to the quick-disconnect coupling at test port 1. Courtesy of Case Corporation.

4. *Hydraulic oil becomes too hot.* Incorrectly set relief valve may be causing a loss of power with the lost energy creating heat in the oil. Leaky piston packing within the cylinder can create localized heating and elevated oil temperature. Inspect oil cooler for damaged fins or restricted air passage during machine operation.

5. *Pump gears, housing, and other parts wear rapidly.* Make certain proper grade and viscosity of oil matches manufacturers recommendations. Inspect for drive alignment.

6. *Relief valves or check valves held partially open.* This can be caused by dirt creating heat and loss of efficiency.

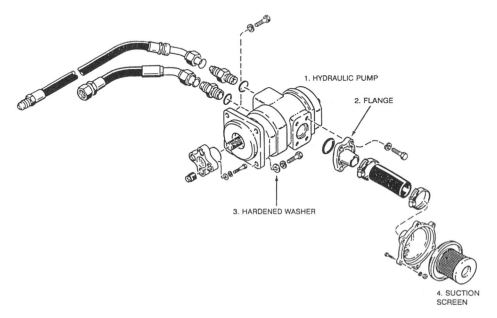

Figure 22.30 Hydraulic pump piping assembly. Courtesy of Case Corporation.

7. *Quick failure of components that have been repaired.* Have oil inspected for quality and conformance to manufacturers specification prior to starting up with new or repaired components.
8. *Cycle times are slow.* The machine does not have enough power.

If the machine has any of these problems, check the hydraulic oil for contamination. A portable filter (Figure 22.31) with a fitting kit (Figure 22.32) provides a convenient method of cleaning the oil. It is also convenient to pump replacement oil from the container into the reservoir to minimize contamination in the fill or refill process.

Types of Contamination

There are two types of contamination: microscopic and visible.

1. Microscopic contamination occurs when very fine particles of foreign material are in suspension in the hydraulic oil. These particles are too small to see or feel. Microscopic contamination can be found by identification of the following problems or by testing in a laboratory. Examples of problems include:

a. Cylinder rods leak.
b. Control valve spools do not return to neutral.
c. The hydraulic oil has high operating temperature.

2. Visible contamination is foreign materials that can be found by sight, touch, or odor. Visible contamination can cause a sudden failure of a component. Examples of visible contamination include:

a. Particles of metal or dirt in the oil.
b. Air in the oil.
c. The oil is dark and thick.
d. The oil has a burned odor.
e. Water in the oil.

Flushing Water from the Hydraulic System

1. Start and run the engine at 1500 rpm.
2. Completely retract the cylinders of all attachments to the machine. *Warning: If retracting the cylinder rods causes the attachment to be raised, block the attachment in place before proceeding to the next step.*
3. Any attachment or part of an attachment that is raised must be supported to prevent the attachment from falling.
4. Loosen and remove the filler cap from the reservoir.
5. Drain the hydraulic oil from the reservoir.
     a. Determine capacity of the reservoir and provide sufficient container capacity to accept the oil that is to be drained.
     b. Remove the drain plug from the bottom of the reservoir.
6. Remove the hydraulic oil filter from the machine.
7. Install a new hydraulic filter.
8. Install the drain plug in the bottom of the reservoir.
9. Fill the reservoir with the recommended quantity of the specified oil.
10. Move each control lever in both directions to release pressure in the circuits.

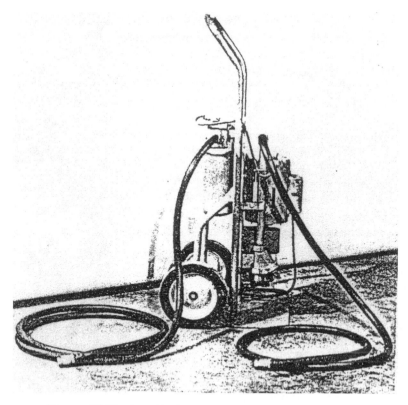

Figure 22.31 Portable filter. Courtesy of Case Corporation.

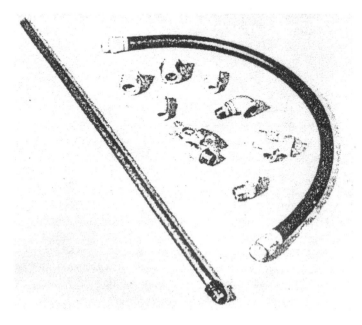

Figure 22.32 Fitting kit for portable filter. Courtesy of Case Corporation.

11.   Disconnect the line from the rod end and closed end of each cylinder.
12.   Be certain all control levers are in the *neutral* position.
13.   Start the engine and run at the engine at low idle. *IMPORTANT: Check the oil level in the hydraulic reservoir while doing step 14.*
14.   Move each control lever in both directions until oil begins to flow from the open line. Hold the control lever in place until clean oil flows from the open line.
15.   Stop the engine.
16.   Connect the line to the closed end of each cylinder.
17.   Start the engine and run the engine at low idle.
18.   Completely extend all cylinders. As the piston rod comes out of the cylinder, oil will be pushed out of the rod end of the cylinder.
19.   Support the loader frame so that the loader frame will stay in the raised position.
20.   Stop the engine.
21.   Connect the lines to the rod end of the cylinders.
22.   Fill the reservoir to the correct level.
23.   Install the filler cap for the reservoir.

Cleaning the Hydraulic System

1.   Prepare the portable filter such as the unit shown in Figure 22.31 or its equivalent by doing the following steps:
  a.   Remove all the hydraulic oil from the inlet and outlet hoses for the portable filter.
  b.   Remove the filter element from the portable filter.
  c.   Remove all hydraulic oil from the portable filter.
  d.   Clean the inside of the housing for the filter element.
  e.   Install a new filter element.
2.   You must know whether the contamination is microscopic or visible (discussed previously).
3.   If the contamination is microscopic:
  a.   Check the maintenance schedule for the machine to learn if the hydraulic oil must be changed. If needed, change the hydraulic oil. Change the hydraulic oil filter.
  b.   Do steps 6 through 38.
4.   If the contamination is visible:
  a.   Change the hydraulic oil and hydraulic oil filter. Remove and clean the hydraulic pump suction screen.
  b.   Do steps 5 through 38.
5.   Check the amount of contamination in the hydraulic system by doing the following steps.
  a.   Disassemble one cylinder in two different circuits. Check for damage to seats, scoring of the cylinder wall, etc. Repair the cylinder as necessary.
  b.   If in your judgment the damage to the cylinder was caused by severe contamination and is not the result of normal wear, it is necessary to remove, clean, and repair valves, pump, lines, cylinders, hydraulic reservoir, etc. in the hydraulic system.
6.   Loosen and remove the filler cap from the reservoir.
7.   Connect a vacuum pump to the filler cap opening. Start the vacuum pump.

8. Loosen and remove the drain plug from the reservoir.
9. There is a shut-off valve within the fitting kit of Figure 22.32 that can be installed in place of the drain plug. Be certain that the valve is closed.
10. Stop the vacuum pump.
11. Connect the inlet hose for the portable filter to the valve that is installed in the hole for the drain plug.
12. Disconnect the vacuum pump from the filler cap opening on the hydraulic reservoir.
13. Insert the outlet hose from the portable filter in the hydraulic reservoir.
14. Open the valve that is installed in the hole for the drain plug.
15. Move the switch for the portable filter to the ON position. Start and run the engine at 1500 rpm.
16. Run the portable filter for at least 10 minutes.
17. Continue to run the portable filter.
18. Increase the engine speed to full throttle. Heat the oil to operating temperature by doing the following steps:
    a. Hold the bucket control lever in the roll-back position for 15 seconds.
    b. Return the bucket control lever to neutral for 30 seconds.
    c. Repeat steps 18a and 18b until the oil in the hydraulic system is at operating temperature.
19. Continue to run the engine at full throttle. Continue to run the portable filter.
20. Operate each hydraulic circuit to completely extend and retract the cylinders. Continue to operate each hydraulic circuit two times, one after the other for 45 minutes.
21. Decrease the engine speed to low idle.
22. Continue to run the portable filter for 10 minutes.
23. Stop the portable filter.
24. Stop the engine.
25. Remove the hose from the hydraulic reservoir.
26. Close the valve that is installed in the hole for the drain plug.
27. Disconnect the inlet hose for the portable filter from the valve.
28. Connect a vacuum pump to the fill cap opening of the hydraulic reservoir.
29. Start the vacuum pump.
30. Remove the valve from the hole for the drain plug.
31. Install the drain plug.
32. Stop the vacuum pump. Disconnect the vacuum pump from the hydraulic reservoir.
33. Install the filter cap.
34. Remove the hydraulic filter element from the machine.
35. Install a new hydraulic filter element on the machine.
36. Check the oil level in the hydraulic reservoir. Add oil as required.
37. Start the engine. Check for oil leakage around the new hydraulic filter.
38. Stop the engine.

Checking Output of Pump with a Flowmeter

Flowmeters are available from several sources. The machine manufacturer will usually recommend an appropriate meter for their machine. The flowmeter in Figure 22.33 is of a

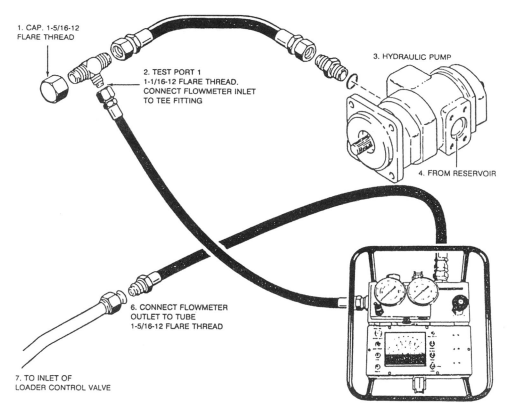

1. CAP. 1-5/16-12 FLARE THREAD

2. TEST PORT 1 1-1/16-12 FLARE THREAD. CONNECT FLOWMETER INLET TO TEE FITTING

3. HYDRAULIC PUMP

4. FROM RESERVOIR

6. CONNECT FLOWMETER OUTLET TO TUBE 1-5/16-12 FLARE THREAD

7. TO INLET OF LOADER CONTROL VALVE

Figure 22.33 Assembly for testing the output of the rear pump section (power steering/loader). Courtesy of Case Corporation.

type recommended by J. I. Case Company for use with their front end loader/backhoe machines.

Checking the Output of the Loader (Rear) Section

1. Park the machine on a level surface. Put the backhoe in the transport position and lower the loader bucket to the floor. Stop the engine.
2. Connect the flowmeter in the circuit (Figure 22.33).
3. Heat the oil to operating temperature by closing the load valve on the flowmeter until the pressure gauge reads 1000 psi (6,895 kPa, 69 bar). Run the engine at full throttle until the temperature reaches 125°F (52°C).
4. Run the engine at 2000 rpm. Open the load valve on the flowmeter. Read the flow gauge. Record the listing.
5. Keep the engine running at 2000 rpm. Slowly close the load valve until the pressure is 2000 psi (13,789 kPa, 138 bar). Read the flow gauge and record the reading.
6. Decrease the engine speed to low idle, then stop the engine.
7. The pump efficiency is found by dividing the flow reading at 2000 psi by the flow reading at 0 psi. The answer multiplied by 100 is the percent efficiency of the pump. If the efficiency is less than 70%, repair or replace the pump.

Checking the Output of the Backhoe or Three-Point Hitch Pump

1. Park the machine on a level surface. Put the backhoe in the transport position and lower the loader bucket to the floor. Stop the engine.
2. Connect the flowmeter into the circuit (Figure 22.34).
3. Heat the oil to operating temperature by closing the load valve on the flowmeter until the pressure gauge reads 1000 psi. Run the engine at full throttle until the temperature reaches 125°F.
4. Run the engine at 2000 rpm. Open the load valve on the flowmeter. Read the flow gauge. Record the flow reading.
5. Keep the engine running at 2000 rpm. Slowly close the load valve until the pressure reaches 2000 psi. Read the flow gauge and record the reading.
6. Decrease the engine speed to low idle, then stop the engine.
7. The pump efficiency is found by dividing the flow reading at 2000 psi by the flow reading at 0 psi. The answer multiplied by 100 is the percent efficiency of the pump. If the efficiency is less than 70%, repair or replace the pump.

TROUBLESHOOTING PROCEDURE FOR PROBLEM IN A SINGLE CIRCUIT

Having studied the troubleshooting procedures for a problem in all circuits, the chart of Figure 22.35 will provide a guide for effectively seeking the individual component(s) that require service. Notes follow relative to the steps in the flow chart.

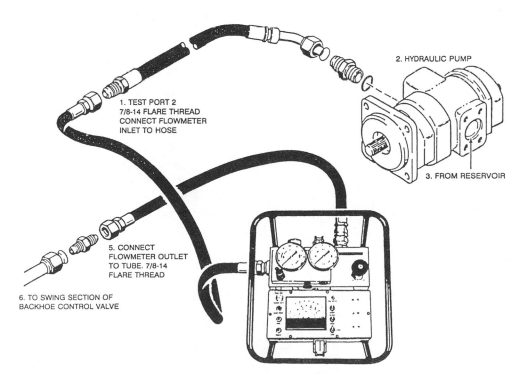

Figure 22.34 Assembly for testing the output of the front pump section (backhoe/three-point hitch). Courtesy of Case Corporation.

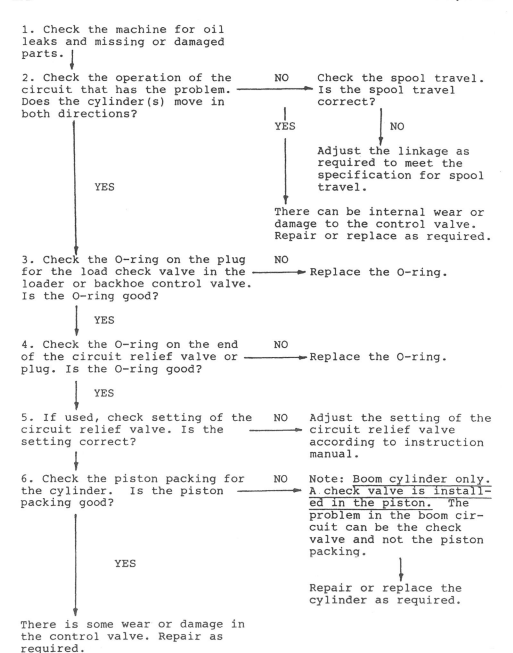

1. Check the machine for oil leaks and missing or damaged parts.

2. Check the operation of the circuit that has the problem. Does the cylinder(s) move in both directions?

NO → Check the spool travel. Is the spool travel correct?

YES | NO

Adjust the linkage as required to meet the specification for spool travel.

There can be internal wear or damage to the control valve. Repair or replace as required.

YES

3. Check the O-ring on the plug for the load check valve in the loader or backhoe control valve. Is the O-ring good?

NO → Replace the O-ring.

YES

4. Check the O-ring on the end of the circuit relief valve or plug. Is the O-ring good?

NO → Replace the O-ring.

YES

5. If used, check setting of the circuit relief valve. Is the setting correct?

NO → Adjust the setting of the circuit relief valve according to instruction manual.

6. Check the piston packing for the cylinder. Is the piston packing good?

NO → Note: Boom cylinder only. A check valve is installed in the piston. The problem in the boom circuit can be the check valve and not the piston packing.

YES

Repair or replace the cylinder as required.

There is some wear or damage in the control valve. Repair as required.

Figure 22.35 Troubleshooting for a problem in a single circuit. Courtesy of Case Corporation.

Step 1. Oil leaks are one indication of a localized problem. Missing or damaged parts can affect valve spool travel and/or alignment of working parts; this requires extra force or rate of movement of machine members and leads to lost time and extra effort. Energy consumption may be unnecessarily increased.

Step 2. Bent or damaged linkage can limit spool travel. Maximum speed of actuator movement will not be attained. Internal leaks in a valve or actuator will limit efficiency of that force and motion function.

Step 3. A leaking load check valve can permit cylinder drift as other functions take priority of the major oil flow.

Step 4. An internal leak can reduce efficiency.

Step 5. Incorrect setting of the circuit relief valve can limit the force capabilities of the actuator(s).

Step 6. External actuator leaks are a nuisance and permit a loss of fluid from the machine requiring refills. Internal leaks reduce actuator efficiency.

SUMMARY

The backhoe priority circuit insures pressurized fluid supply to power steering whenever the engine is operating. The front end loader provides adequate force and motion functions with the inboard pump. The three-point hitch provides adequate force and motion functions with the outboard pump. The backhoe functions with supply from both the inboard and outboard pump after the power steering priority valve has supplied potential needs for the steering circuit. The steering circuit is not normally functional when the backhoe is functional. Banked valves are employed to minimize circuit piping.

A working knowledge of ISO symbols is an essential for troubleshooting the hydraulic circuit. Study of the sectional view of the components aids in understanding the power flow through the circuits.

External leaks, excessive heat, excessive noise, and erratic motion provide a key to the detection of malfunctioning circuit components. Use of these key factors minimizes the time involved in the troubleshooting function.

23

Typical Hydraulic Press Circuits

The typical force and motion pattern associated with a hydraulic press for forming, bending, coining, and shear operations entails a rapid approach at low force levels; a high force is needed for punch, shear, or forming operations. At the completion of the major work function a rapid return to the desired position completes the cycle.

Two basic circuits are involved to develop these force and motion requirements. The first is a regenerative circuit. The flow from the rod end of the major cylinder is joined with the supply flow to the head or blind end of the cylinder. Thus the cylinder moves at the rate needed to displace the rod area. The force that can be developed is that of the rod area (Figure 23.1a). The objective in the circuit of Figure 23.1a is to provide equal rate of movement in each direction of travel rather than a force and speed variable. The addition of suitable valving will provide a change from the regenerative function (Figure 23.1a) to a conventional function at the desired portion of the machine cycle to provide the increased force at the end of the stroke. At this desired point in the machine cycle the flow from the rod end will be directed to tank so that full force can be developed by the piston pushing against the rod and load (Figure 23.1b).

SHIFT FROM RAPID ADVANCE TO FULL FORCE AT A SLOWER SPEED

To initiate the closing mode solenoid S2 is energized; this actuates the pilot valve of directional valve (D, Figure 23.1b). Major fluid flow is directed to the head end of the cylinder. Flow from the rod end is directed through check valve E to join with the major pump flow. Valve A is biased to the closed position. The pilot line to valve B is connected to the tank. It is protected from shock loads or high back-pressure by check valve C. When major resistance is encountered by the cylinder rod in its forward movement it is reflected to the pilot of valve A. When the set value of valve A is reached the flow from the rod end is directed to tank and the full piston area has maximum force. In the return mode solenoid S1 is energized, directing pump flow to the rod end of the cylinder. Flow to this line provides a pressure signal ahead of check valve C to pilot valve B. This pilot signal causes valve B to

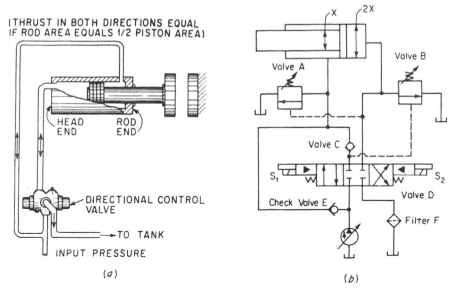

Figure 23.1 Differential circuits. (a) Thrust in both directions will be of equal value assuming that the rod area equals one-half of the piston area. Piston speed will be identical in each direction of travel when supplied with a uniform flow. (b) Circuit with valving to automatically change from a differential action to a full force conventional function. Regenerative flow through check valve E will be diverted through valve A.

move to the open position. Part of the flow from the head end can pass through valve B in the return stroke to minimize the pressure drop and attain maximum return speed.

JACK RAMS AND THE PREFILL FUNCTION

The second hydraulic press circuit employs one or more jack rams to move the main ram(s) during a rapid movement of the press. At the desired position the main ram is pressurized to provide the desired force value (Figure 23.2).

Flow to the main ram(s) is through a small high-pressure line and/or through a prefill valve. A prefill valve is a pilot operated check valve that can be opened by atmospheric pressure pushing the oil through the free-flow passage past the spring-biased closure poppet and/or the opening function can be assisted by an auxiliary pilot piston. The auxiliary pilot piston insures full travel of the poppet from the seat in the desired cycle pattern.

The main ram of Figure 23.2 employs the lip of the ram to aid gravity during the return function by pressurized fluid directed to connection A. The jack ram moves the main ram and lower platen of the press of Figure 23.2 in the up movement. The main ram functions as a pump during this upward movement with fluid flowing through poppet C from the reservoir. As the press closes and pressure increases valve C (Figure 23.3) is shifted directing flow through check valve B to the main ram. The vacuum created by the main ram is negated by the flow through check valve B and full force is available on the total area of the main ram to force the rising platen against the fixed platen. The effective

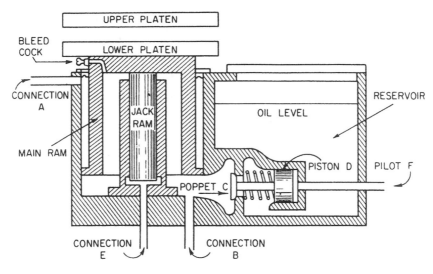

Figure 23.2 Sectional view of hydraulic press with moving lower platen and rigid upper platen. Lower platen retracts with pressurized fluid to lip of main ram plus gravity.

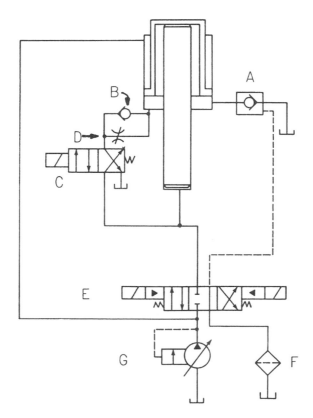

Figure 23.3 Typical circuit for the press of Figure 23.2 with the use of solenoid actuated directional control valves.

area of the jack ram and main ram and the pressure level determines the force available to close and hold the platens in the work mode.

Pressure at connection A and pilot F (Figure 23.2) with connections E and B to the tank permits retraction of the main and jack ram to the open position of the platens. Orifice D (Figure 23.3) releases high pressure oil at a controlled rate prior to the opening of piloted prefill valve A. Releasing the high pressure oil at a controlled rate is required to minimize shock as the stressed press structure returns to the relaxed position.

MOVING UPPER PLATEN

The ram assembly and reservoir can be above the press platens. The operation is similar with the exception that a counterbalance valve is normally installed in the lines to the jack ram(s) to avoid platen drop when the press platens are in the open position.

THE PREFILL VALVE

The prefill valve may be an integral part of the press assembly (Figure 23.2). The prefill valve may be a standardized component whether fitted into the pipeline, manifolded to the main ram, or assembled into the ram or reservoir in a cartridge format. Figure 23.4 shows a typical prefill valve designed to be manifolded into the cavity above the main ram. It is fitted with a small diameter poppet assembly in the center of the major prefill check poppet. The relationship to the orifice and parallel check valve is carefully calculated so that change from the prefill function to the major force mode is assured.

The small poppet designed into the major poppet assembly to releases high pressure oil prior to the movement of the major closure as the press changes to the opening mode; this is designed appropriate to the pressure level that is anticipated and the cycle rate of the press. Figure 23.5 illustrates a typical prefill valve and its position within the flow path from the reservoir in the crown of the press to the main ram. It employs a throttle valve to adjust

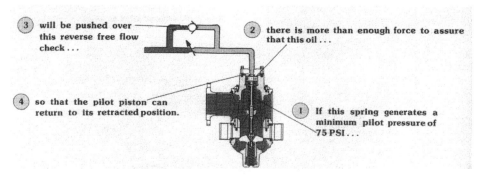

Figure 23.4 The prefill pilot assembly is designed to insure adequate bias spring force to return oil past the flow rate control valve structure. The prefill valve is designed to be inserted into the crown of the press above and as an integral part of the main ram. A flange connection is provided to connect the valve via a pipeline to the associated reservoir normally positioned above the press. Courtesy of Rexroth Corporation.

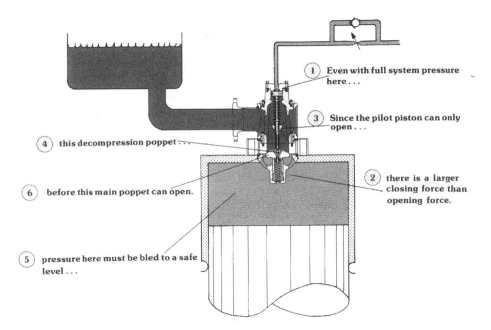

(1) **Even with full system pressure here . . .**

(3) **Since the pilot piston can only open . . .**

(4) **this decompression poppet . . .**

(2) **there is a larger closing force than opening force.**

(6) **before this main poppet can open.**

(5) **pressure here must be bled to a safe level . . .**

Figure 23.5 Area relationships of the prefill valve are designed to insure adequate decompression prior to initiation of major flows at high pressure. The prefill valve of Figure 23.4 is shown in position with the connection to the reservoir. Courtesy of Rexroth Corporation.

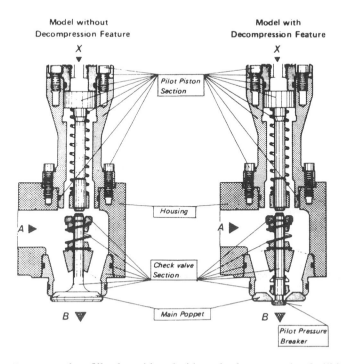

Model without Decompression Feature

Model with Decompression Feature

X

X

Pilot Piston Section

Housing

A ▶

A ▶

Check valve Section

B ▼

B ▼

Main Poppet

Pilot Pressure Breaker

Figure 23.6 Pilot operated prefill valve with and without the decompression facilities. The valve to left is directly actuated; the valve to right incorporates a decompression feature. Courtesy of Rexroth Corporation.

the decompression rate of the oil at high pressure. A check valve is provided to permit the pilot piston to return to its retracted position as the press cycle calls for high force. Some press circuits employ auxiliary circuits to provide the decompression function. Thus the decompression feature becomes an option in the choice of prefill valves. The prefill valve to the left in Figure 23.6 is designed without a decompression feature. The integral decompression poppet concentrically located in the valve at the right in Figure 23.6 functions mechanically as pilot pressure is applied at port X. As in the previous examples relaxation of high pressure through this small passage minimizes shock and permits relaxation of the press structure.

Figure 23.7 illustrates the total circuit for a press designed so that the upper platen is the moving member. As illustrated the function is rapidly advance the upper platen with the main ram functioning as a pump. Two jack rams are employed to provide the desired motion pattern. The major ram functions as a pump during part of the cycle and (when the prefill has closed) as an additional area for creating the desired tonnage or force pushing the platens together during the pressing function.

Figure 23.8 illustrates the opening function. Flow through the counterbalance valve passes the major control structure via the integral check valve to the rod area of jack or kicker cylinders.

TROUBLESHOOTING

Platen Drift

Trouble may consist of platen drift when press is running or idle (Figure 23.9). Undetected platen drift can damage the die or perhaps injure the operator.

Step 1. The rate of platen drift can be fast or slow.

Step 2. A leak may be hidden within the press structure, so the shrouding may have to be removed. Excessive oil around the machine may provide a clue.

Step 3. The counterbalance adjustment may be too low if a heavier die has been placed in machine. Foreign matter in the counterbalance valve may have changed the set force values.

Step 4. The free-flow poppet within the counterbalance valve may be damaged or held from seat with foreign matter. *Close the press or block platen before working on counterbalance valve.*

Step 5. Other moving parts within the counterbalance valve may be damaged requiring replacement of the complete valve assembly. *Close press or block platen before working on counterbalance valve.*

Step 6. A damaged jack ram may cause drift. *Close press or block platen before working on jack rams.*

Slow Pressure Buildup

The press may close at normal speed but encounter excessive dwell as the prefill valve closes (Figure 23.10). Pressure builds slowly up to the set value required to perform the desired function.

Step 1. A restriction in the pilot line to the prefill valve may cause slow closing. Damage in the prefill pilot may also slow closing action and/or not permit complete closure.

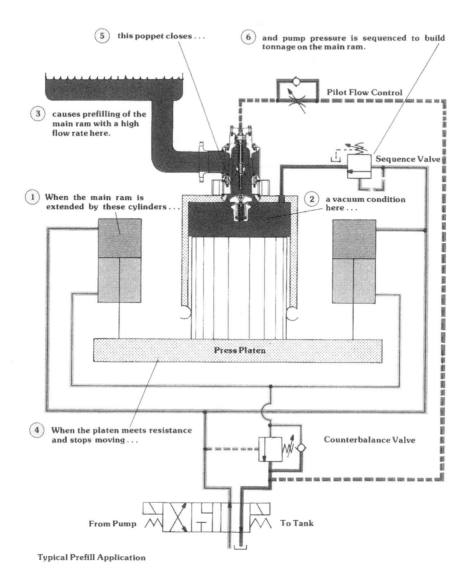

Figure 23.7 Total hydraulic circuit illustrating flow during the closing of the platens ready for the pressing cycle. Note the controlled flow through the counterbalance valve as it is piloted open from the line to the jack rams and sequence valve. The counterbalance valve insures a closing rate commensurate with the rate of fluid supply to the jack rams. Full pressure is available to the major diameter of the main ram as resistance is encountered. Flow through the sequence valve as the resistance level reaches the value of the sequence valve setting negates the vacuum condition at the prefill valve. As the prefill valve closes full tonnage can be exerted by the main and jack rams. Courtesy of Rexroth Corporation.

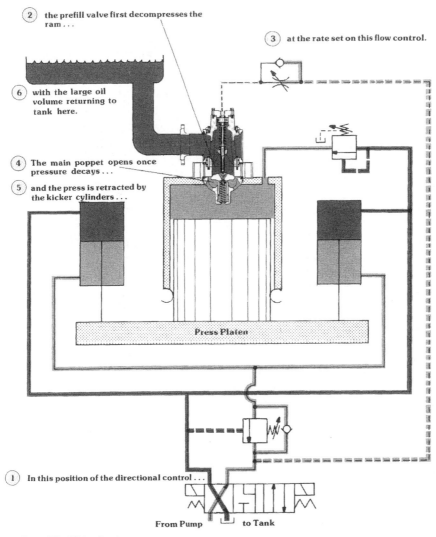

(2) the prefill valve first decompresses the ram . . .

(3) at the rate set on this flow control.

(6) with the large oil volume returning to tank here.

(4) The main poppet opens once pressure decays . . .

(5) and the press is retracted by the kicker cylinders . . .

Press Platen

(1) In this position of the directional control . . .

From Pump   to Tank

Typical Prefill Application

Figure 23.8 Total hydraulic circuit illustrating flow during the opening of the platens to complete the press cycle. Flow passes through the integral check valve in the counterbalance valve. Pilot pressure for the decompression function is taken from a point adjacent to the counterbalance valve. Courtesy of Rexroth Corporation.

Step 2. Foreign matter also can prevent full closure of the prefill valve and is probably lodged in the pilot actuator. Major flow through the prefill valve usually washes the seat clean. A pressure breaker-type assembly may be jammed causing enough leakage for slow pressure build-up and therefore the pump to run hot.

Step 3. Overheated oil may oxidize and deteriorate. Have samples tested and replace if needed.

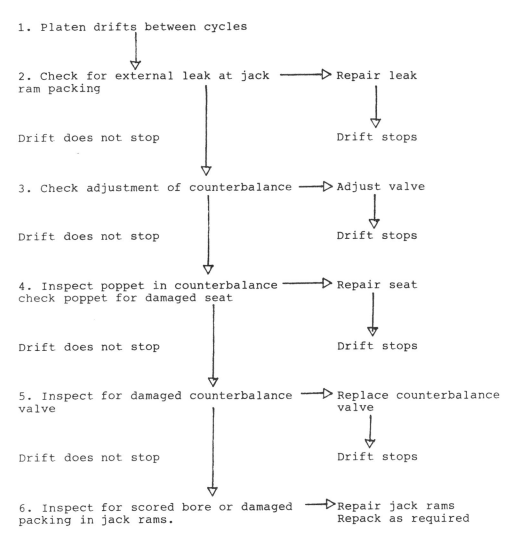

Figure 23.9   Troubleshooting for drift when upper platen moves.

Step 4.   The pump compensator may need adjusting to respond at the proper rate when prefill closes.

Step 5.   A worn pump will create heat because of slippage oil. If additional cooling does not increase the viscosity of oil to compensate for pump wear, a repair or replacement is needed.

Excessive Shock

Excessive shock may become obvious gradually as press functions in normal working pattern. Several potential causes need to be addressed in a random pattern, but step 1 (Figure

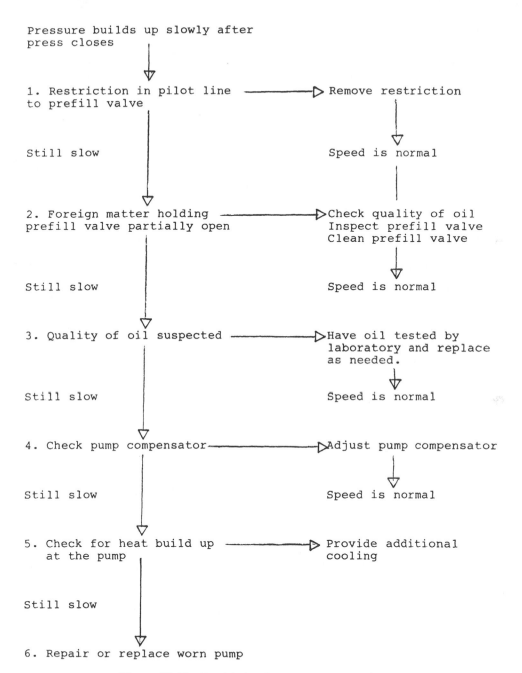

Pressure builds up slowly after
press closes

1. Restriction in pilot line ——————▷ Remove restriction
to prefill valve

Still slow   Speed is normal

2. Foreign matter holding ——————▷ Check quality of oil
prefill valve partially open   Inspect prefill valve
                                    Clean prefill valve

Still slow   Speed is normal

3. Quality of oil suspected ——————▷ Have oil tested by
                                    laboratory and replace
                                    as needed.

Still slow   Speed is normal

4. Check pump compensator——————▷ Adjust pump compensator

Still slow   Speed is normal

5. Check for heat build up ——————▷ Provide additional
     at the pump   cooling

Still slow

6. Repair or replace worn pump

Figure 23.10 Troubleshooting slow pressure buildup.

23.11) is a likely potential cause. If the shock appears in a short time frame, steps 2 through 5 may be most likely suspects.

Step 1. An adjustable restriction is often provided in the line to the prefill pilot to control the opening speed and as a result the rate at which the captive high pressure fluid is decompressed. An increased restriction due to foreign matter will increase press cycle time.

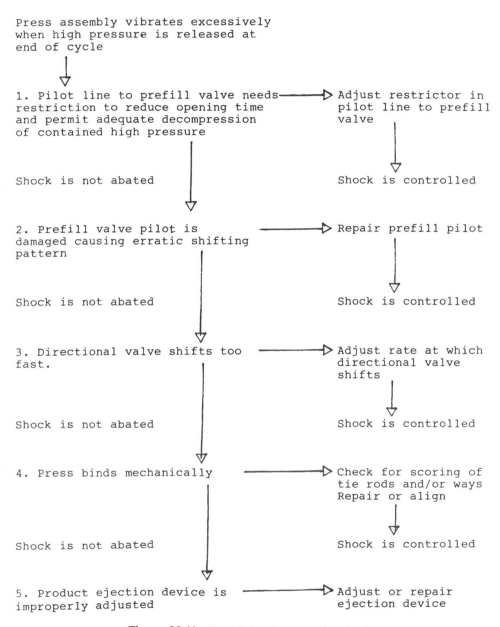

Figure 23.11 Troubleshooting excessive shock.

Incorrect adjustment to a fuller opening can increase both cycle time and shock potential. This adjustment is usually the first to be checked for shock problems.

Step 2. Foreign matter restricting the prefill pilot in an erratic pattern can cause intermittent shock conditions. Inspection and repair may need to be accompanied by replacing or reconditioning the oil.

Step 3. A directional valve supplying oil to the prefill pilot may be affected by hot or cold oil, thereby changing the shifting rate. The controlled pilot pressure to the directional valve via a reducing valve may improve valve shifting pattern and reduce shock resulting from erratic supply to the prefill valve.

Step 4. Mechanical binding of the press is often difficult to detect until major damage is incurred. Check for scoring of tie rods and/or ways. Make certain way and tie-rod lubrication systems are functioning properly.

Step 5. Product ejection devices can create shock loads because of improper lubrication of dies, poor alignment, or mechanical wear.

SUMMARY

Transmission of power via a contained and pressurized liquid is unquestionably the most effective medium for this difficult and essential force and motion task associated with coining, forming, compressing, and shearing.

Shock in hydraulic presses can be anticipated by the nature of the force and motion requirements. The magnitude of the shock will be related to the cycle time. An increase in production by means of faster cycle time can increase shock potential. Appropriate decompression of captive oil and relaxation of the press structure between each cycle can reduce the shock to an acceptable level. A suitable compromise is an attainable goal. Maintaining the established performance level with acceptable shock can insure a long machine life with adequate product quality.

24

Troubleshooting Agricultural Machinery

Agricultural machinery involves many force and motion functions, some requiring brute strength and a high degree of precision, and others requiring agility, fast reversals, and repetitive accuracy. There are many similarities between the functions needed for ground preparation, cultivation, and harvesting, and those involved in machinery for municipal utility maintenance, new construction, mining, timber harvesting, and many other comparable industrial manufacturing activities. Dependability is a key requirement for agricultural machinery. The local weather and planting and harvest schedules are rarely negotiable. Thus a new facet of maintenance and careful design enters the fluid power force and motion functions as employed in machinery for agricultural activities.

FLUID POWER CIRCUITS

Both open and closed circuits may be employed in the force and motion functions for a typical agricultural machine. While each circuit may function independently, the circuits may share a common reservoir and fluid conditioning systems. The reservoir is a significant part of the open hydraulic circuit. All flow throughout the circuit comes from the reservoir, through the pump and control valves, to the actuator, and then back to the reservoir. The reservoir serves in a somewhat different capacity for a closed hydrostatic circuit. It becomes the source of makeup fluid and a facility for storing and conditioning the hydraulic fluid.

Hydrostatic Circuit Charge Pump

A charge pump of relatively small capacity, usually manifolded to or within the major pump structure, pumps oil from the reservoir (Figure 24.1). The output from the charge pump is directed through check valves to the low pressure side of the hydrostatic circuit. This supply of pressurized oil is employed to make up for oil used in the lubrication and cooling of the motor, power pump, and actuators. This low pressure oil is also employed to assist in returning the pistons within the hydrostatic pump after the completion of the power stroke. Lubrication oil, sometimes referred to as slip flow, is directed from the pump and

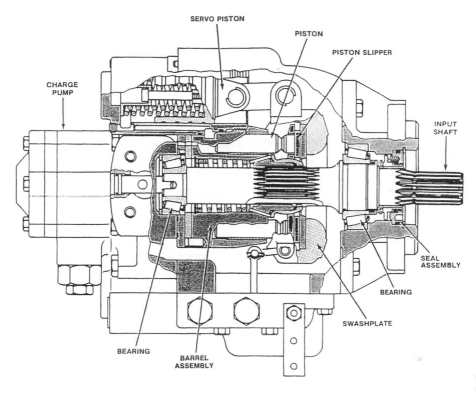

Figure 24.1   The charge pump for a hydrostatic transmission is manifolded to the variable displacement piston pump with a tang drive as an extension of the input shaft. Courtesy of Case Corporation; Case 7-7.

actuator casing to the reservoir. It is pressurized, cleaned, and cooled within the make-up pump circuit prior to returning to the reservoir or being introduced back into the hydrostatic circuit.

Hydrostatic Drive

All traction power for a machine can be via pressurized fluid powering one or more hydraulic motors. The pump supplying pressurized fluid to the motor provides the directional function and rate of flow which determine motor speed and vehicle travel rate. Pressure level within the circuit is a result of resistance encountered by the machine in the functional work mode. The motor is an integral part of a wheel or tracked structure with the power applied as desired by the designer and user. Independent motors can be added to the basic drive circuit for traction at the desired point when operating in difficult terrain.

The major advantage of the hydrostatic drive is the ease of control. Infinite speed control from the start-up mode to maximum designed speed rate is a feature of the variable displacement pump assembly. A change of flow rate, direction, or fluid pressure related to resistance to movement can be accommodated by the hydrostatic transmission to provide any desired ground speed and energy application appropriate to the task to be performed.

Another obvious advantage is flexibility of component locations. The motor and pump can be placed within the machine remote from each other with all power moving via the contained pressurized fluid lines. This gives a flexibility of design not found in mechanical systems. Another less obvious advantage is the elimination of many gears, shafts belts, and pulleys that may constitute a hazard to the operator and service personnel. The integral lubrication supplied by the hydraulic oil is another basic advantage to the hydrostatic transmission circuit.

Axial Flow Combine

The J. I. Case Axial Flow Combine incorporates a hydrostatic front wheel traction drive system that is also available with an auxiliary rear wheel drive assembly. Other force and motion requirements which are a part of the harvesting function are powered by pressurized oil from spur gear pumps.

The hydrostatic drive shares a common reservoir with the rest of the hydraulic system. It is located at the upper left side of the machine, just in front of the hydraulic pump. The recommended fluid is Hy-Tran Plus hydraulic fluid. The amount of oil in the reservoir depends on the combine model. Check the operator's manual for correct fill level.

Two filtration systems were employed in the hydraulic systems. One is inside the reservoir itself (Figure 24.2). This element has a 40-μ rating and does not contain a by-pass valve. This element, if used, should be changed whenever the hydraulic oil is drained. Always drain the reservoir to change this filter. The service interval is once a year or every 1000 hours. The external filtration system consists of two external spin-on filters (Figure 24.3). These elements are 95% efficient at a 10-μ rating. The service interval is 1000 hours for the filters and once per year for the oil. The J. I. Case Rice machine uses a pressure filter for the hydrostatic charge circuit (Figure 24.4), while all other machines use a suction filter for the charge circuit.

Integral Charge Pump

The displacement of the charge pump can be changed by increasing or decreasing the thickness of the gerotor-type gear set appropriate to the functions served by the charge pump. The charge pump serves four basic functions:

1. It provides oil flow and pressure to the directional valve, servo pistons, and, if equipped, the two-speed motor pressure response valve, two-speed motor servo pistons, and the Equa-Trac valve for switching the rear wheel drive on or off.
2. It makes up for oil lost due to leakage from the hydrostatic closed loop, keeping the system primed.
3. It provides cooled, filtered fluid for temperature control and flushing. The amount of cooler flow will equal the output of the charge pump on machines that do not have a power guide axle. Those with power guide axles will have slightly less cooler flow than the charge pump rating.
4. Provides flow, under pressure, for maintaining back-pressure on the motor and pump pistons.

Charge Pressure Relief Valve

A pilot operated relief valve located next to the charge pump relieves the charge fluid into the main pump casing when the hydrostatic transmission is in neutral and the closed loop is full of oil. This valve is factory preset at 220–240 psi at a flow rate appropriate to the

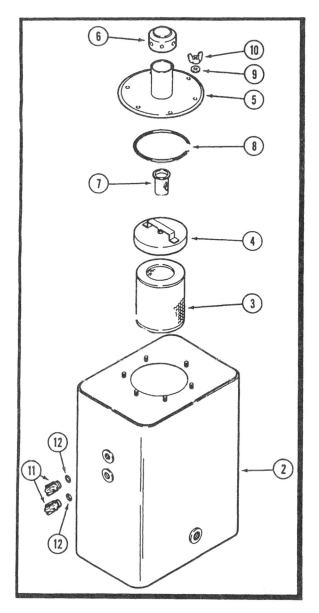

2.   Reservoir Tank
3.   Filter
4.   Filter Cover
5.   Tank Cover
6.   Breather Cap
7.   Strainer
8.   O-Ring
9.   Washer
10. Wing Nut
11. Sight Glass
12. O-Ring

Figure 24.2   Reservoir with internal filter. Courtesy of Case Corporation; Case 7-3.

intended service. It is stamped with the numbers 022 to identify its pressure setting (022 indicates 220 psi).

Directional Control Valve for Hydrostatic Transmission

This valve is mounted to the main hydrostatic pump and is operated by a single cable connected to the propulsion lever in the cab. The valve provides a metered flow of charge

RESERVOIR

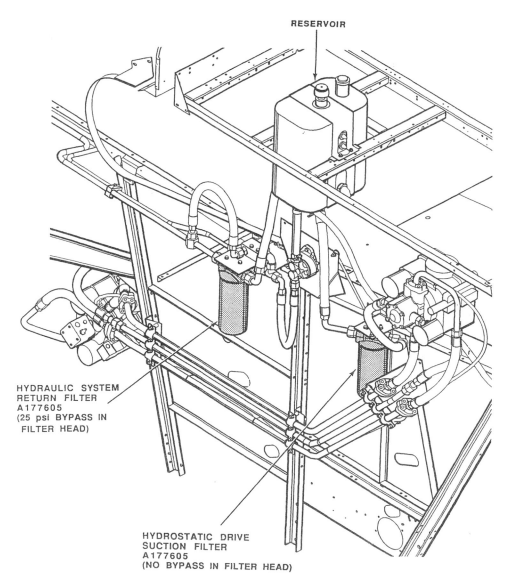

HYDRAULIC SYSTEM
RETURN FILTER
A177605
(25 psi BYPASS IN
FILTER HEAD)

HYDROSTATIC DRIVE
SUCTION FILTER
A177605
(NO BYPASS IN FILTER HEAD)

Figure 24.3 Spin-on filter locations. Courtesy of Case Corporation; Case 7-4.

system oil to the forward or reverse servo pistons. It is a closed center valve. Oil flow is only present when the operator wants to make the machine move. Once speed is established, only enough oil will flow through this valve to maintain the set speed. This compensates for normal servo system leakage.

High Pressure Lines

Lines used to connect the hydrostatic pump to the hydrostatic motor (or motors) are made of either a complete steel tubing assembly or a combination of steel and steel-braided rubber

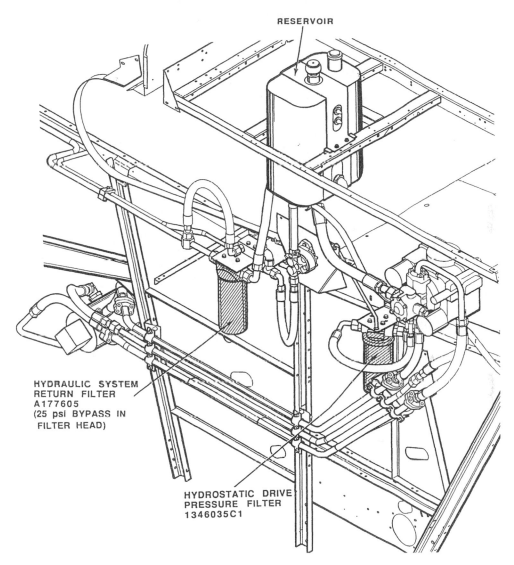

RESERVOIR

HYDRAULIC SYSTEM
RETURN FILTER
A177605
(25 psi BYPASS IN
FILTER HEAD)

HYDROSTATIC DRIVE
PRESSURE FILTER
1346035C1

Figure 24.4 Hydrostatic drive pressure filter for rice harvester. Courtesy of Case Corporation; Case 7-5.

lines. One line is for forward drive pressure while the other is for reverse. The smaller of the three lines running from the pump to the motor is for circulation and cooling flow and is a low pressure assembly.

Single-Speed Hydrostatic Motor

A fixed displacement axial piston motor is bolted directly to the combine transmission and drives it through a splined coupler. The motor has essentially the same basic internal components as the pump except that the swash plate is fixed at an 18 degree angle and there are no servo pistons.

Two-Speed Hydrostatic Motor

This motor differs from the fixed displacement motor because the swash plate angle is variable from 15 to 18 degrees. This motor has a pressure response valve mounted to it; these valves control the motor servo pistons, which in turn set the swash plate from 15 to 18 degrees. The servo pistons in this unit are not spring centered.

High Pressure Relief Valves

Mounted to the motor is a valve block (Figure 24.5). Two of the components of this block are the high pressure relief valves: one for the forward drive pressure and one for reverse pressure. The valves are pilot operated and dump the high pressure flow of the closed loop to the low pressure side of the loop if the relief pressure is reached. They also provide pilot fluid to the foot-n-inch valve. Pressure values vary with different combines from approximately 5500 to 6700 psi.

Foot-n-Inch Valve

The left pedal in the cab is connected to a pressure control valve located under the cab. When the pedal is in the relaxed (not depressed) position the rod-and-spring assembly

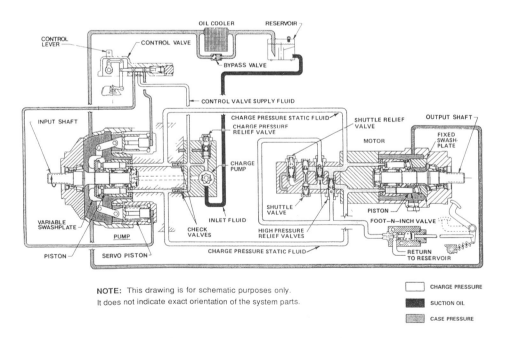

Figure 24.5 Axial piston variable displacement pump and fixed displacement motor in neutral position. (Note: This drawing is for schematic purposes only; it does not indicate the exact orientation of the system parts.) Courtesy of Case Corporation; Case 7-18.

provide the needed bias to establish the maximum desired high pressure in the hydrostatic drive system. As the pedal is depressed it relaxes pilot pressure from the high pressure relief valve (Figure 24.5). This valve serves as the pilot poppet for both high pressure relief valves. It is the first piece to move in the high pressure relief system if maximum drive pressure is reached. It is also where the maximum drive pressure for the combine is set. As previously noted this valve also can be manually operated by depressing the left pedal in the cab. As the foot-n-inch pedal is gradually depressed the maximum system pressure decreases until the vehicle stops. The foot-n-inch circuit is shown symbolically in Figure 24.6. *Caution*: When the foot-n-inch pedal is depressed, the machine will free wheel. Coasting downhill will damage the hydrostatic unit since the motor will be rotating with no high pressure lubrication. Also, never tow the combine with the transmission in gear.

Internal Pressure Override Control

An internal pressure override system is incorporated into the hydrostatic drive system of certain models. The devices are occasionally referred to as horsepower limiters. The objective is to reduce the pump stroke at a predetermined pressure value to prevent heat build-up due to operation at very high drive pressures for long periods of time.

The combine is controlled through the directional control valve. In addition to this control, is the override valve (Figure 24.7) that is connected in series before the directional control valve. If the system high pressure exceeds the setting of the pressure override control, the pressure available for servo piston control over the swashplate is gradually closed off. This allows the swashplate of the pump to modulate toward neutral and prevents the system from dumping fluid over the main relief valve in the motor valve block, thereby preventing heating of the oil.

In this condition, the machine will slow down or eventually stop if the drive pressure is not reduced. The machine may seem sluggish to the operator. To regain response, the propulsion lever should be moved to neutral, and the transmission should be run in a lower gear to reduce the amount of drive pressure necessary to move the machine through the field. Typical setting for the override valve is 6000 psi, and is shim adjustable. The foot-n-inch valve is set at 6700 psi as a backup system and manual dump.

Shuttle Relief Valve

The action of the shuttle relief valve is shown in Figure 24.6. Pressure from the high pressure side of the loop causes the shuttle to provide a flow from the low pressure side of the loop through the shuttle relief valve to the pumps, case, cooler, and then back to the reservoir. This relief valve is identical to the charge relief valve but the setting is lower. Typical setting is 160–190 psi.

Oil Cooler

A fin-and-coil type of unit is located in the clean air chute after the rotary screen. It cools the oil that has passed through the hydrostatic system as flow over the charge or shuttle relief valves, and oil that has come from the closed loop as normal leakage and lubrication. Combines with 12 gpm or more flow from the charge pumps have two oil coolers and larger lines to handle the extra flow with minimum restriction. An oil cooler bypass consists of a ball-and-spring type of pressure relief valve located near the cooler. This works as a result of differential pressure across the oil cooler. It protects the cooler and low pressure circuit

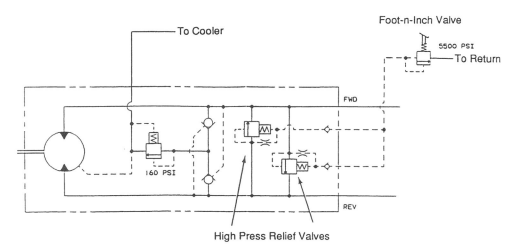

Figure 24.6 Foot-n-inch circuit shown with ISO symbols. Courtesy of Case Corporation; Case 7-32.

from over pressurization due to added restriction from cold oil. It is set for a 15 psi differential.

Directional Control Valve for Swashplate Control

The directional control valve is shown in section and symbolically in Figure 24.8 (note the centering spring). In the neutral position the swashplate control pistons are both connected to tank. Pressure from the make-up pump is blocked. The swashplate feedback insures accurate positioning.

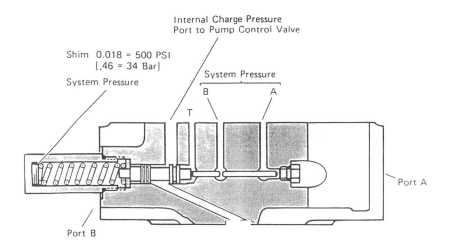

Figure 24.7 Internal pressure override control valve. Courtesy of Case Corporation; Case 7-37.

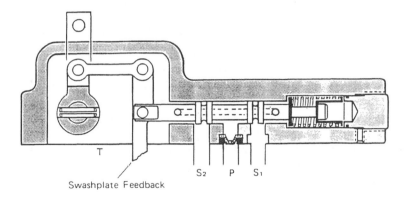

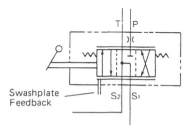

Figure 24.8 Directional control valve for swashplate position control. Courtesy of Case Corporation; Case 7-23.

Hydrostatic Drive

Neutral

Neutral is considered to be when there is no output flow being generated by the variable displacement pump, as its shaft, internal components, and charge pump are being driven. With no flow available, the motor delivers no output to the transmission (Figure 24.5).

The control valve in the neutral position blocks all flow to the servo pistons. The servo pistons, in turn, are centered in the neutral position by springs. In the neutral position, the variable swashplate is vertical and no pumping action occurs from the rotating group.

Oil is drawn from the reservoir by the charge pump; this pump then supplies oil to the control valve, where it is blocked in neutral. Oil from the charge pump opens the check valves in the pump-end cover, thereby priming the system (including the pump pistons and high and low pressure lines). The oil also primes the motor and its high pressure relief valves. The shuttle valve is spring centered and blocks flow to the shuttle relief valve at this time. Oil is also exposed to the foot-n-inch valve where its path is also blocked.

The charge pump builds pressure in the system until it reaches the charge pressure relief valve setting (220–240 psi). The oil pressure opens the charge relief valve and the oil is expelled into the pump case. This oil fills the motor case via the single low pressure line running from the pump to the motor. The pump case then overfills and the oil proceeds to the cooler, then back to the reservoir. The cooler flow equals the output of the charge pump. In neutral, cooler flow circulates to the charge pump inlet.

Forward

When the operator moves the propulsion lever in the cab, the directional control valve spool moves off center, exposing the port to the forward servo piston to charge pressure (Figure 24.9). Charge fluid is routed through a control response orifice, past the spool land, and through a drilled passage to the forward servo piston. The forward servo piston is pushed out of its cylinder, causing the swashplate to tilt. As the swashplate tilts, the feedback linkage, which is connected to the spool of the directional valve, moves. This movement recenters the spool and prevents any more oil from going to the forward piston once the desired ground speed is reached. The reverse servo piston is pushed into its cylinder forcing its fluid into the pump case through the opposite port of the directional control valve. The reverse servo piston return springs are under compression at this time and are trying to return the swashplate to a vertical position. The servo pistons have no seal rings on them, so there will be slight leakage of servo pressure to the pump case. As the pressure in the forward servo piston decreases, the springs in the reverse servo piston begin to straighten the swashplate. If this condition is allowed to proceed, the combine would eventually slow down and stop. To prevent this, the feedback linkage will automatically shift the directional control valve spool slightly off center and replace any leakage from the pressurized servo piston. Therefore, the combine will maintain the set speed as long as the propulsion lever stays in place.

The swashplate is mounted on trunnion bearings. As the swashplate tilts, it changes

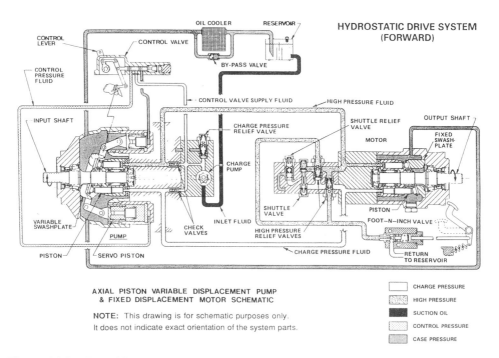

Figure 24.9 Control lever moved to the forward position. Control pressure fluid is directed to the servo piston to position the swashplate for appropriate supply of fluid to the motor to cause forward motion. (Note: This drawing is for schematic purposes only; it does not indicate the exact orientation of the system parts.) Courtesy of Case Corporation; Case 7-25.

the amount of oil being pumped; the greater the angle of tilt, the larger the volume of fluid is displaced. Since the motor has fixed displacement, it will rotate faster to accept the larger volume of oil.

The high pressure oil exits the rotating block in the pump through kidney-shaped holes. This fluid seats the forward check valve, blocking off charge fluid on the forward side of the closed loop. The fluid is directed to the motor via the heavy high pressure tubes. As the fluid enters the motor, it is exposed to the pistons on half of the motor block. These pistons will be forced out of their bores causing them to slide down the fixed motor swashplate. This causes the motor block to rotate, which of course, turns the motor output shaft.

High pressure fluid is also directed into the motor valve block. In the valve block, fluid enters the inside of the forward high pressure relief valve, washes the ball off its seat, and is directed to the foot-n-inch valve by an external line. Another external line directs the fluid to the reverse high pressure relief valve where the ball is seated by the fluid. Within the valve block, high pressure fluid goes around the reverse high pressure relief valve to one end of the shuttle valve. The shuttle valve moves away from the high pressure fluid; this opens a passage in the opposite side of the closed loop. This passage allows the return side of the loop to be exposed to the shuttle relief valve. This valve is set at 160–190 psi and maintains this pressure in the return side of the closed loop. Pressure opening the valve allows fluid to be expelled into the motor case causing flow to occur around the working rotating group. This flow dissipates heat from the motor. *Never* switch the charge relief valve with shuttle relief valve as this will result in an overheated motor. Once the fluid overfills the motor case, it goes to the pump case. The fluid is used to cool the rotating components of the pump, overfills the pump, and goes to the oil cooler. After cooling, the fluid is returned to the reservoir.

Back at the charge pump, the reverse side check valve is moved off its seat by charge pressure. This allows the charge pump to replace all the oil lost from the closed loop due to normal high pressure leakage and the quantity that went past the shuttle relief valve. If the total amount of leakage from the high pressure loop exceeds the output of the charge pump, the system will no longer function.

Reverse

Reverse operation is the same as forward with the following changes:

1. The control lever is moved to the opposite of neutral. This moves the spool in the directional control valve in the opposite direction from forward. The fluid flow and pressure is now applied to the opposite servo piston. The variable swashplate now rotates in the opposite direction. This in turn reverses the flow from the high pressure pump (Figure 24.10).

2. High pressure fluid enters the opposite side of the motor cylinder barrel assembly and causes the motor to rotate in the opposite direction.

3. The reverse high pressure relief valve sends oil to the foot-n-inch valve and is now the protecting valve in the drive system.

4. The high pressure fluid moves the shuttle valve to the other end of its cavity. This allows the low pressure side of the loop access to the shuttle relief valve and creates a cooling flow for the motor and pump.

5. The charge pump flow enters the circuit by way of the opposite check valve on this low pressure side of the loop. *Important*: The hydrostatic drive is a fail-safe system. If

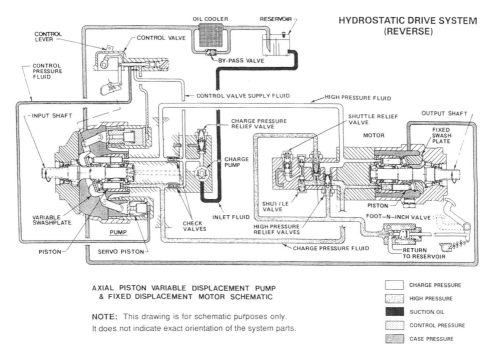

Figure 24.10 Control lever moved to the reverse position. Control pressure fluid is directed to the servo piston to position the swashplate for appropriate supply of fluid to the motor to cause reverse motion of the machine. (Note: This drawing is for schematic purposes only; it does not indicate the exact orientation of the system parts.) Courtesy of Case Corporation; Case 7-29.

there is a loss of fluid, the charge pump loses pressure thereby allowing the servo springs to return the swashplate to neutral.

Two-Speed Hydrostatic Motor

The two-speed hydrostatic drive motor has a moveable swashplate that is servo operated within a range of 15–18 degrees. An electrical solenoid switch located in the operator's cab next to the fuse panel provides a HI and LO position. Motor components consist of a variable displacement motor, pressure response valve, shifting valve and valve block containing high pressure relief valves, shuttle valve, and shuttle relief valve. The variable motor is very similar to the variable pump with the exception that the charge pump and centering springs on the servo pistons are not present, and swashplate mechanical stops limit movement from 18 to 15 degrees.

The main feature of the two-speed hydrostatic motor is to provide a 17% increase in ground speed when operating in first or second gear in the HI range. When the transmission is shifted to the third HI range, the two-speed motor is shifted back to low range by a lockout switch located in the shifter console.

Operation in the HI range automatically shifts back to the LO range for ground speeds providing increased torque when adverse field conditions prevail. Operation in the HI range provides faster speeds that are desirable in many field conditions.

Pressure Response Valve

The pressure response valve offers several options. In low speed mode the solenoid is deenergized and the conditions outlined in Figure 24.11 will be encountered. In the high speed position and with drive pressure under 3100 psi, the swashplate in the motor is moved to the 15 degree tilt and minimum displacement. If the drive pressure increases beyond 3100 psi the control piston overrides the existing condition and causes the swashplate to return to the 18 degree position, thereby creating the maximum torque value.

Thus when operating in the HI range (swashplate at 15 degrees) and in adverse field conditions (mud or soft ground), drive pressure will increase. Normal drive pressures will be 2000–2300 psi on firm ground. The pressure response valve located on the two-speed motor monitors increases in the drive pressure. The pressure response valve is a variable motor control that selects the operating range for the present operating conditions.

The pressure response valve is factory set to operate within a range of 3100–3500 psi. When drive pressure exceeds 3100 psi in the HI range, deceleration starts. When drive

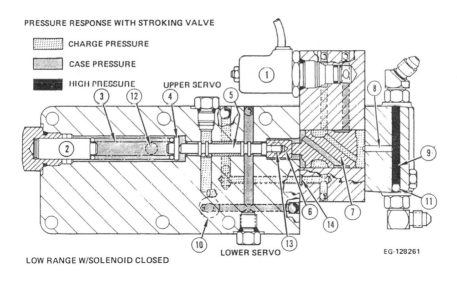

PRESSURE RESPONSE WITH STROKING VALVE

CHARGE PRESSURE

CASE PRESSURE

HIGH PRESSURE

UPPER SERVO

LOW RANGE W/SOLENOID CLOSED

LOWER SERVO

EG-128261

1 - Solenoid Assembly	8 - Needle Roller
2 - Spring Retainer Plug	9 - Drive Pressure
3 - Spring	10-Orifice Plate
4 - Spring Retainer	11-Check Ball
5 - Pressure Response Spool	12-Case Drain
6 - Spool Cap	13-Case Drain
7 - Control Piston	14-Orifice

Figure 24.11 Two speed switch and control valve assembly. The solenoid valve assembly (1) is deenergized and blocks the oil going to the case drain (13). This allows the charge pressure to increase behind the control piston (7), overcoming the pressure response spool spring (3). The pressure response spool then moves, exposing the charge pressure to the upper servo cylinder port changing the swashplate angle to 18 degrees. This decreases the ground speed and increases torque. Courtesy of Case Corporation; Case 7-43.

pressures reach 3500 psi, the two-speed hydrostatic motor is now operating in the LO range and the swashplate angle is at 18 degrees (maximum stroke).

The shifting valve operates in conjunction with the pressure response valve. This valve gives the operator the selection between the HI and LO ranges. The shifting valve is a solenoid operated valve that directs the pressure response valve to set the motor at maximum or minimum displacement. It is electrically operated and requires a 12 volt DC in-line (5 amp.). *Important*: A switch has been added to the shift gate to prohibit the engagement of the HI range in 3rd gear. This was done for reasons of safe highway operation.

TESTING

Purging the System

Whenever the system is opened for a filter change, servicing, or replacement of components, the following procedure should be followed before the machine is moved. All hydraulic lines should be connected in the sequence recommended in the service manual and torqued to the specified values.

Place the transmission in neutral and start the engine. Set the engine speed at high idle. Move the control lever approximately one fifth of the way forward. Allow the system to operate for 4–5 minutes. Move the control lever back through neutral and then one fifth of the way in reverse. Allow the system to operate for 4–5 minutes. Return the control to neutral. Repeat this sequence several times. This procedure will usually dislodge any trapped foreign material and move it back to the reservoir.

Test Procedures

Hydrostatic Drive

Typically all tests are conducted with fluid temperature at 150°F. Before testing the system the following checks should be made.

1. Fluid level in reservoir (with feeder house lowered).
2. Condition of the hydraulic oil. Does it contain water, contamination or give evidence of excessive heat?
3. Condition of the hydrostatic control cable. Is it stiff or hard to move? Check the position of the propulsion control lever in each quadrant: neutral, forward, and reverse. Check the position of the actuating arm on the pump actuating bracket assembly.
    a. It should be positioned parallel with the pump.
    b. The actuating arm in the bracket should align with the pump actuating arm.
    c. Is the actuating arm worn excessively?

After these items have been checked, testing can begin.

Pressure Levels

1. Install a 10,000-psi gauge at the high pressure side of foot-n-inch valve (Figure 24.12). The drive pressure of both forward and reverse can be observed at this location. (*Note*: Tee the gauge into the line; do not dead head.)

Figure 24.12 10,000 psi gauge installed at foot-n-inch valve. Courtesy of Case Corporation; Case 7-53A.

2. Install a 600-psi gauge in the charge pump test port on the hydrostatic pump (Figure 24.13). If the unit is equipped with a two-speed motor, attach the gauge to the tee on the hydrostatic pump and in the charge pressure line going to the two speed motor. *Important*: During the test, the machine must be allowed to creep. The motor must rotate (drive wheels moving) for accurate checks of the system. Caution! Be certain no one is in the vicinity of the wheels or in front of the machine.

3. System charge pressure. Place transmission in 3rd gear. Start engine and set at high idle. Observe charge pressure (in neutral it should be 220–240 psi at the charge pump relief valve).

Figure 24.13 600 psi gauge installed on hydrostatic pump assembly. Courtesy of Case Corporation; Case 7-53B.

    a.   Charge pressure forward/reverse. Move propulsion control level into forward part of quadrant. Charge pressure should be 160–190 psi at the shuttle relief valve located in motor. *Note*: It may take a slight amount of drive pressure build-up to get the charge relief valves to switch. Move propulsion control lever into reverse part of quadrant; the same response should be seen as in forward.

    b.   Drive pressure. Restrict the machine's movement using the brakes or other safe means, then shift the transmission to third gear. Move the propulsion control lever forward in the quadrant until maximum drive pressure is achieved. Allow the wheels to creep during the test, but restrict the machine's movement using the brakes or other safe means; then shift transmission to third gear. Move the propulsion control lever rearward in the quadrant until maximum drive pressure is achieved. Allow the wheels to creep during the test.

    4.   Foot-n-inch valve leakage. Check rate by removing the *return line* and observing the leakage rate at 4500 psi (25 cc or 1 oz/min maximum). For safety, connect a hose to the fitting where the return line was removed and allow the fluid to flow into a small clean container. Do not stand in front of the combine during the test.

    5.   Servo cylinder pressure test. Install two 300-psi gauges in the directional control valve (Figure 24.14). Set parking brake and shift transmission to neutral. Start engine and set at high idle. Place the directional control valve in full forward or reverse (move the propulsion lever all the way forward or reverse). Observe pressure readings—forward 150–170 psi, reverse 50–70 psi. (The reverse pressure will be lower because the control cable limits the amount of movement to the directional control valve. This limits reverse ground speed for safety purposes.) If no pressure is read in one direction, check for leaks externally and internally on the directional control valve.

    6.   Two-speed hydrostatic motor. Install a 600-psi gauge in the lower servo test port of the two-speed motor control valve. (See Figure 24.15 for test port location.) With the range switch set on HI, engage the hydrostatic drive and observe the gauge readings at the

Figure 24.14   600 psi gauges installed in directional control valve. Courtesy of Case Corporation; Case 7-54.

Figure 24.15 60 psi gauge installed in motor case drain port. Courtesy of Case Corporation; Case 7-55.

pressure response valve. At a drive pressure of 3100 psi and below, the gauge should read 160–190 psi; at a drive pressure above 3500 psi, the gauge should read 20–40 psi (case pressure). If the specified pressure cannot be obtained, the variable motor must be removed and the response valve repaired or replaced.

Contamination

Failures that are determined to be caused by contamination or can be determined to have introduced contamination in the form of metal particles or other such impurities into the system should be treated as catastrophic failures. Because of the nature of the hydrostatic drive system, any contaminants left in the system upon a replacement of a failed system component can be expected to cause a failure again as the contamination accelerates wear on system parts. In order to avoid this problem the following steps should be taken: change the pump, motor, oil cooler, and bypass valve; flush all lines and the reservoir; change the filter; finally, replace the hydraulic fluid. Following this procedure will greatly reduce the chance of any material being left in the system from the previous failure. While this is an expensive process, it is much less expensive in time and materials than having the machine fail again while in the field, or damaging new components from the old failure. The system must be *clean*.

TROUBLESHOOTING

The Hydrostatic Drive

The test readings provide a key to the source of a malfunction in the hydrostatic drive system. Possible causes, test readings and remedies are listed in Table 24.1.

Rear Wheel Drive

The rear wheel drive system consists of two cam lobe wheel motors, control valve, and high and low pressure lines. A switch on the right-hand console activates a solenoid used to turn the unit on and off. The high pressure fluid to drive the rear wheels comes directly from the hydrostatic pump. One line is forward and the other is for reverse. The control valve is located on the rear axle. This control valve contains a solenoid, selector spool, and flow dividers to provide positive traction to the rear wheels. As the fluid enters the wheel motors, it is used to move the pistons in their bores causing the hubs to rotate.

The rear wheel drive axle is available as a factory installed option. This system provides additional driving force to the machine to get through tough field conditions. Both rear wheels are driven by hydraulic motors. This system receives its fluid directly from the engine-driven hydrostatic pump. The high pressure lines provide fluid to the wheel motors through a control valve. The control valve utilizes an electrically operated solenoid valve that is engaged or disengaged by a switch in the cab.

This power assist operates as a function of the standard hydrostatic drive system. The standard hydrostatic system utilizes one pump and one motor, whereas the rear wheel drive axle system utilizes one pump and three motors. The two wheel motors are types of cam lobe units with fixed displacement. The amount of wheel torque or drive force obtained from these units is strictly related to hydrostatic system pressure: the higher the system pressure, the greater the driving effort will be.

It is very important to maintain traction to the front wheels. If a wheel loses traction, the system pressure will lower since the spinning wheel requires less pressure to rotate. The oil in the lines will take the path of least resistance; therefore the amount of power assist received from the rear wheels will be reduced because of the lower drive pressure. Shifting to a high gear (if possible) will reduce the amount of torque and increase the drive pressure at the front wheels to stop wheel slip. Also, lightly touching the brakes may help.

There are significant advantages to using the rear wheel drive at all times when working in the field. Benefits include lower system operating pressure, improved steering, and lower hydraulic oil temperature because of lower overall drive pressure.

When transporting the machine to or from the work site and/or when resistance to movement is low, the rear wheel switch should be turned off. The cam lobe motors automatically free-wheel when the switch is turned off because these wheels do not have drive hubs to disengage. There is no limit to the distance the machine can be driven at one time. *Note*: The front and rear wheels do not require any type of device to synchronize them. The system pressure and fluid drive do this automatically. Figure 24.16 illustrates the circuit associated with the rear wheel drive system.

The Pilot System for Actuating the Rear Wheel Motors

The cartridge-type solenoid valve employed to engage the two rear drive motors is located in the valve assembly that is identified as the Equa-Trac valve. A low pressure shuttle assembly is also located in the Equa-Trac valve (Figure 24.17). The low pressure shuttle provides charge fluid to the solenoid valve to engage and disengage the rear wheel drive system. The charge fluid is directed to either end of the selector spool to accomplish this. Charge fluid is available from the return side of the hydrostatic system. The spacer (2) always holds one ball off its seat. This assures pilot oil for the system even if the main hydrostatic system is in neutral. When the combine is driven forward, the forward side becomes high pressure (drive pressure) and the reverse side is low pressure (charge pressure). The higher pressure on the forward side will seat its ball (3), which will push the spacer (2) to the right unseating the ball on the reverse side. The charge fluid on the reverse side is directed past the ball (3) and spacer to the drilled passage (4) leading to the solenoid valve (solenoid valve input supply).

With the combine in reverse, the reverse side becomes high pressure and the forward side is low pressure. The reverse ball (3) seats and the spacer moves to unseat the opposite ball. This now directs charge fluid past the ball and spacer (4) leading to the solenoid valve. The solenoid valve now determines to which end of the selector spool the oil is directed.

When the switch in the cab is depressed to the off position, the current to the solenoid valve is disconnected. The spring-loaded spool moves away from the solenoid coil, shutting off the fluid supply to the selector spool (Figure 24.18). The return spring and charge pressure moves the selector spool to the right (disengaged position). The selector spool closes off both the forward and reverse high pressure ports (Figure 24.19). The rear wheels will now free-wheel. As the wheels move, the motor pistons are pushed back into their bores. This fluid from the pistons is directed to the case pressure return. A drilled passage in the selector spool directs the fluid through an external line, to the solenoid valve, and on to case drain.

Engaging the Rear Wheels

Charge pressure is directed to the selector spool by the solenoid valve (Figure 24.20). As the switch is moved to energize the solenoid to engage the rear wheel motors, there is a momentary, but important, intermediate position of the selector spool (1) (Figure 24.21). Prior to full high pressure engagement, the intermediate position allows for charge pressure to hold the pistons against the cam lobes. This is accomplished by feathering lands and a drilled passage in the selector spool entering the cored passages (forward and reverse) of the valve. This pressurization of the pistons at charge pressure reduces the high pressure spike that can occur when pressurizing with drive pressure.

The selector spool movement is controlled by the 0.020 in. orifice (7) on the left-hand side of the selector spool (1). This orifice will meter the fluid out of the cavity, thus allowing the low pressure shuttle fluid to fill the circuit prior to drive pressure entering. Using charge pressure to engage the motor pistons reduces the shock load exerted on the pistons and the motor housing when turning the unit on.

When going forward, high pressure fluid enters the forward port, goes past the selector spool land, and enters the center hole of the forward flow divider (Figure 24.22). Fluid exits each end of the flow divider and is directed towards the wheel motors through the check ball seats (3). Pressure will be the same at both wheel motors and provides positraction to the wheels.

If fluid pressure should differ from one motor to the other, the valve will compensate for it. The lower pressure motor will be the path of least resistance, so it will get more fluid

Table 24.1 Troubleshooting the Hydrostatic Drive (Cases 7-60, 7-61, 7-62, 7-63, 7-64)

Possible cause	Test reading	Remedy
1. Transmission will not operate in either direction		
A. System low on fluid	Very low or zero charge pressure.	1. Locate and fix leaks causing loss of fluid. 2. Check fluid level in reservoir and replenish if necessary with clean Case IH Hy-Tran Plus Fluid.
B. Faulty control linkage to pump	Charge pressure and case pressure normal. Case pressure with control level moved forward or backward.	1. Check the entire linkage to make sure it is connected and is free to operate as it should. 2. Be sure the linkage is adjusted to the neutral position of the pump control arm.
C. Disconnected internal control valve linkage	Charge pressure and case pressure normal, but pump control lever moves freely with no resistance when disconnected from the external control linkage. No servo pressure can be produced.	Check for broken or missing parts. Replace the control valve as needed. The pump must be disassembled and flushed to remove missing or broken parts.
D. Plugged control valve orifice	Charge pressure and case pressure normal. Pump control lever moves with normal resistance and centers to neutral due to centering spring. No servo pressure.	Remove the control valve and clean the orifice.
E. Charge pump suction line or filter plugged	Low charge pressure and low case pressure.	Clean the suction line. Replace the filter. Make sure suction line is not collapsing.
F. Pump or motor drive coupling disconnected	—	Check drive line.
G. Charge pressure relief valve in charge pump or motor valve manifold stuck open or damaged	Low or zero charge pressure. If the charge pressure is low at neutral the faulty valve is in the charge pump. If charge pressure is low when the control valve is moved in either direction, the faulty valve is in the valve block of the motor.	Replace faulty valve.
H. Charge pump drive shaft or key broken	Zero charge pressure.	Replace charge pump.
I. Internal damage to pump or motor or both	Low or zero charge pressure. Charge pressure may also fluctuate rapidly or fall to or near zero when attempting to stroke transmission control. Units are noisy. Pieces or flakes of brass in the filter reservoir. Cannot make drive pressure.	Replace the pump and motor. One unit or both are damaged and the generated contamination has been spread throughout the system. Flush entire circuit with clean solvent and blow dry with clean air before installation of a new system. Replace the oil cooler and bypass.

2. *Transmission operates in one direction only*		
A. Faulty control mechanism	Refer to 1.B	Refer to 1.B
B. Faulty check valve	System high pressure loss in one direction only.	Replace the faulty check valve. Make sure no pieces of the old valve are in the pump.
C. Damaged high pressure relief valve	Lower than normal or loss of high pressure in one direction only. *Note:* Switch high pressure valves and test to see if the transmission operates properly in the previously inoperative direction.	Replace the valve.
D. Shuttle valve jammed open	Lower than normal or loss of high pressure in one direction only.	Remove shuttle valve and repair if possible. Replace motor valve block if necessary.
E. Control valve spool sticking in one direction	Control handle at operator console will not stroke in one direction.	1. Check cable and all linkage first. 2. Replace the control valve assembly.
3. *Sluggish performance*		
A. Control orifice partially blocked	Refer to 1.D.	Refer to 1.D.
B. Air in system	Erratic pressure readings on all gauges.	1. Check for inlet line air leaks. 2. Check reservoir fluid level.
C. Internal damage to pump or motor or both	Refer to 1.I.	Refer to 1.I.
4. *System noisy*		
A. Air in system	Refer to 3.B.	Refer to 3.B.
B. System tubing or pump or motor not properly installed	Vibration noise.	1. Check tube clamping. 2. Check pump and motor mounting.
5. *Difficulty in finding neutral (may be impossible to find)*		
A. Control valve out of adjustment	—	Replace the control valve.
B. Control linkage faulty	—	Disconnect the control linkage at the control valve lever. If the system returns to neutral, replace or adjust the control linkage.
C. Servo sleeve out of adjustment	—	Remove pump and rezero swash plate following Eaton service information or return the pump for factory readjustment.

Table 24.1 Continued

Possible cause	Test reading	Remedy
6. *Variable displacement motor will not change displacement*		
A. Control pressure line from motor valve manifold to pump housing plugged	No control (charge) pressure to motor response valve.	Replace or repair the line.
B. Plugged control orifice	No pressure at motor servo pistons.	Clean or replace orifice valve.
C. Solenoid not functioning properly	Can only get pressure to one servo.	Check electrical circuit. Repair or replace solenoid.
D. Stuck main shifting spool	Can only get pressure to one servo.	Repair or replace valve.
7. *Transmission operating hot (motor case temperature above 180°F)*		
A. Dirty air cooler	All gauge readings normal.	Replace air cooler and the bypass.
B. Oil cooler bypass valve faulty	Refer to 7.A.	Repair or replace bypass valve.
C. Case drain line improperly installed	Fluctuating case pressure could be high or low.	Correct the installation.
D. Low fluid	Low or zero charge pressure; low or zero drive pressure	Fill reservoir to proper level.
E. Excessive operation of transmission at high pressure relief valve setting	Refer to 7.A. Also, high pressure gauge shows operation of transmission to be occurring at high pressure relief valve setting during a specific duty cycle.	1. Shift into a lower gear. 2. Refer to the operator's manual for proper machine operation.
F. Filter or suction line partially plugged	High vacuum, low charge pressure, and low case pressure.	Replace filter and clean suction line. Make sure suction line is not collapsing.
G. Excessive internal leakage	High pressure readings are low in one or both transmission output directions. All other pressures normal. Charge pressure lower than normal; may drop off severely at high pressure. Also refer to 1.I.	1. Check the foot-n-inch valve poppet, seat, and spring. 2. Replace high pressure relief valve(s) that are faulty. 3. Refer to 1.I.

8. Troubleshooting guide for the rear wheel drive axle

Problem	Possible cause	Remedy
1. Performance is sluggish	A. Operating with the transmission in low gear	This system is more effective in a higher gear; shift up.
	B. Forward and reverse high pressure hoses installed incorrectly.	Refer to the service manual for correct installation of hoses.
	C. Incorrect high pressure relief setting.	Refer to the service manual for proper specification.
	D. Excessive high pressure leakage from: 1) Wheel motor(s), 2) Equa-Trac II valve.	Refer to pressure tests in the service manual. Repair as necessary.
	E. Charge pressure low.	Refer to pressure tests in the service manual. Repair as necessary.
2. The unit will operate in forward but not reverse or operates in reverse and not in forward		Refer to pressure tests in the service manual. Repair as necessary.
3. One wheel is dragging or locked up.	A. Incorrectly installed high pressure hoses at the wheel motor.	Refer to the service manual for correct installation of hoses.
	B. Wheel motor failed.	With the system turned off and the tire off the ground, rotate the tire by hand. The wheel should free wheel.
	C. Stuck flow divider in Equa-Trac II valve.	
4. Rear wheel not providing driving force.	A. Blown fuse.	Replace fuse.
	B. Failed switch.	Replace switch.
	C. Broken wire between the switch and the solenoid.	Repair wire.
	D. Loose wire at the solenoid.	Tighten.
	E. Solenoid coil failed (resistance 7–9 ohms).	Replace coil.
	F. Solenoid valve spool stuck or binding.	Replace solenoid valve.
	G. Main spool in Equa-Trac II valve not shifting.	Refer to test procedure in the service manual.
	H. Plugged 0.020-in. orifice on left side of selector spool.	

Before attempting to troubleshoot, be sure to check fluid level, replace the filter, check the suction hose, and correct any external leaks. Warm to fluid until it is at least 150°F before testing the system. Be certain there is no air in the system.

Source: Case Corporation.

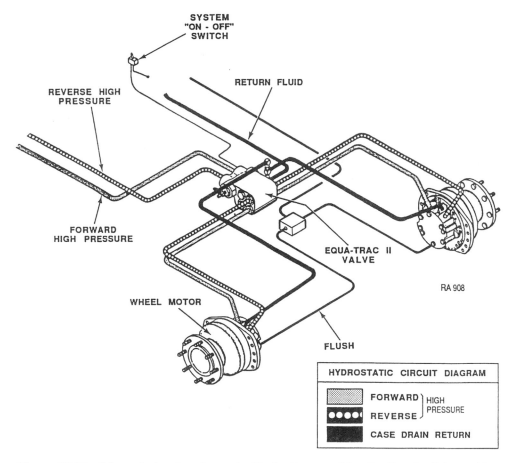

Figure 24.16 Schematic drawing of rear wheel hydraulic motor drive. For "roading" the machine, switch the rear wheel drive off. The cam lobe motors automatically free-wheel when the switch is turned off. The system does not have drive hubs to disengage. There is no limit to the distance the machine can be driven at one time. Note: The front and rear wheels do not require any type of device to synchronize them. The system pressure and fluid drive do this automatically. Courtesy of Case Corporation; Case 7-68.

(and rotate faster). However, the higher pressure will be felt on the opposite side of the flow divider. The high pressure will push the flow divider towards the low pressure side, and limit the amount of fluid going to that side. This insures that the wheel with traction gets drive pressure. When pressure requirements equalize, the flow divider will shift back to the center position. If the flow divider is pushed far enough to one side that it shuts off the flow to that side, the orifice enters the channel and provides some oil flow. This eliminates the chance of completely closing off one side.

The return fluid from the wheel motors enters the Equa-Trac valve at the reverse ports when in the forward drive position (Figure 24.23). Thus fluid enters the check ball seats and is directed to each side of the flow dividers. The fluid enters the flow divider and exits out the center hole. The fluid now flows past the selector spool land and back to the main hydrostatic pump.

The check balls are in the system to prevent an excessive amount of back-pressure in

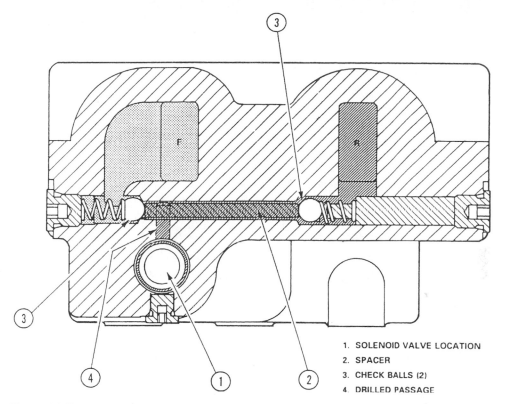

1. SOLENOID VALVE LOCATION
2. SPACER
3. CHECK BALLS (2)
4. DRILLED PASSAGE

Figure 24.17 Low pressure shuttle segment of Equa-Trac valve shown in cross section. Courtesy of Case Corporation; Case 7-78.

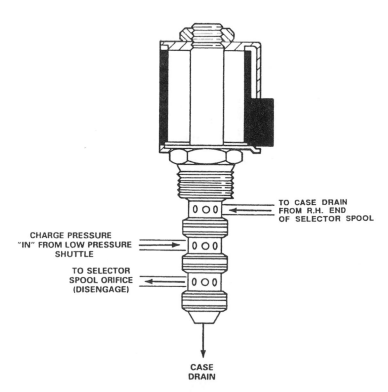

TO CASE DRAIN
FROM R.H. END
OF SELECTOR SPOOL

CHARGE PRESSURE
"IN" FROM LOW PRESSURE
SHUTTLE

TO SELECTOR
SPOOL ORIFICE
(DISENGAGE)

CASE
DRAIN

Figure 24.18 Solenoid valve port locations. Courtesy of Case Corporation; Case 7-80.

OFF POSITION

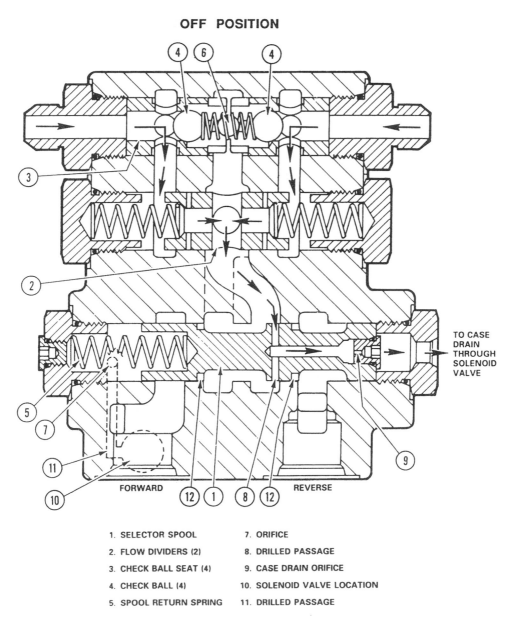

1. SELECTOR SPOOL
2. FLOW DIVIDERS (2)
3. CHECK BALL SEAT (4)
4. CHECK BALL (4)
5. SPOOL RETURN SPRING

7. ORIFICE
8. DRILLED PASSAGE
9. CASE DRAIN ORIFICE
10. SOLENOID VALVE LOCATION
11. DRILLED PASSAGE

Figure 24.19 The selector spool (1) closes the forward and reverse high pressure ports. Courtesy of Case Corporation; Case 7-81.

the return circuit. Should pressure increase beyond the normal 160–190 psi, the check ball will move off its seat, diverting the fluid directly to the center hole of the flow divider.

Reverse Flow

Reverse flow through the Equa-Trac valve is the exact opposite of forward. High pressure enters the reverse port, goes past the spool land, and enters the center hole of the reverse

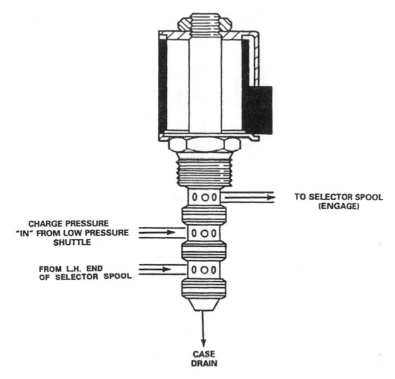

TO SELECTOR SPOOL
(ENGAGE)

CHARGE PRESSURE
"IN" FROM LOW PRESSURE
SHUTTLE

FROM L.H. END
OF SELECTOR SPOOL

CASE
DRAIN

Figure 24.20 Flow conditions with solenoid valve energized. Courtesy of Case Corporation; Case 7-82.

flow divider. The fluid goes out each end of the flow divider, past the check balls, and out to the motors. The return fluid exits the valve through the forward port and flows to the pump.

Split Flow Divider Spools

Early Equa-Trac valves employed the one-piece flow divider assembly. Later models employed a split spool design (Figure 24.24). The split flow divider spools operate in a similar manner to the one piece design. Oil pressure is maintained to the wheel that has traction. However, the split flow divider spools will also limit the maximum amount of flow available to the rear axle motors. If both wheels should begin to spin, the split flow divider will separate because of oil pressure acting on a .328-in. orifice in each half of the spool. As each half moves apart, this limits each rear motor to approximately 18 gpm. The rear axle is therefore limited to 36 total gpm. With the one-piece spool design, if both rear wheels are spinning, the spool may stay centered and full hydrostatic pump flow could go to the rear motors.

Cam Lobe Motor

The cam lobe motors (Figure 24.25) are low speed, high torque motors that directly drive the wheels. Each motor contains 10 piston assemblies (2) located in a cylinder block assembly (6). The cylinder block is located inside of the cam ring (1). Each piston has a follower (3) that rides up and down the cam ring lobes as the unit rotates. The cylinder

INTERMEDIATE POSITION

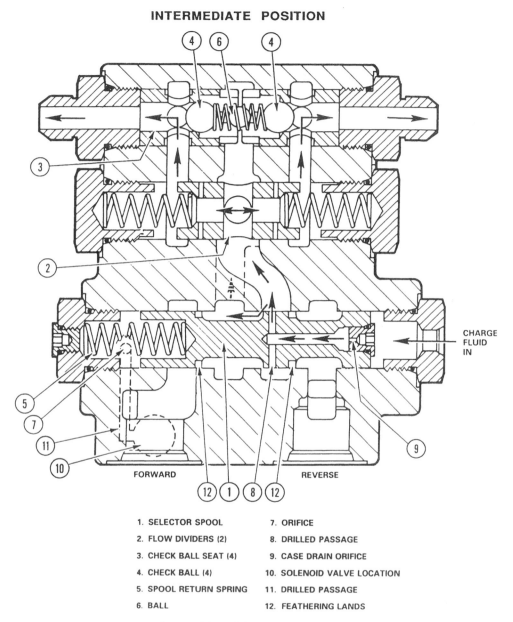

1. SELECTOR SPOOL	7. ORIFICE
2. FLOW DIVIDERS (2)	8. DRILLED PASSAGE
3. CHECK BALL SEAT (4)	9. CASE DRAIN ORIFICE
4. CHECK BALL (4)	10. SOLENOID VALVE LOCATION
5. SPOOL RETURN SPRING	11. DRILLED PASSAGE
6. BALL	12. FEATHERING LANDS

Figure 24.21 The intermediate position of the Equa-Trac valve. Courtesy of Case Corporation; Case 7-83.

block is splined to accept the splined shaft of the output shaft (wheel shaft). An oil distribution block (4) is bolted to the cam ring. This provides communication from the rotating cylinder to the inlet and outlet passages on the wheel motor housing.

Pressurized oil from the Equa-Trac valve enters the inlet port on the forward side of the wheel motors. This oil is sent into the oil distribution blocks of the motors. The oil

ON POSITION

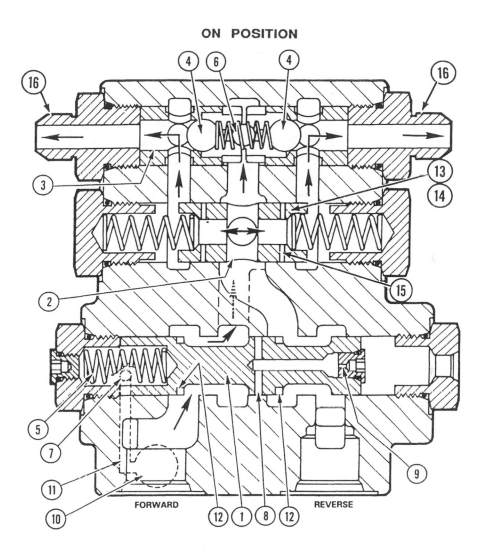

1. SELECTOR SPOOL
2. FLOW DIVIDERS (2)
3. CHECK BALL SEAT (4)
4. CHECK BALL (4)
5. SPOOL RETURN SPRING
6. BALL
7. ORIFICE
8. DRILLED PASSAGE
9. CASE DRAIN ORIFICE
10. SOLENOID VALVE LOCATION
11. DRILLED PASSAGE
12. FEATHERING LANDS
13. FLOW DIVIDER (FORWARD)
14. FLOW DIVIDER (REVERSE)
15. ORIFICE
16. FORWARD DRIVE PORTS
17. REVERSE DRIVE PORTS

Figure 24.22   The Equa-Trac valve fully shifted to the ON position. Courtesy of Case Corporation; Case 7-85.

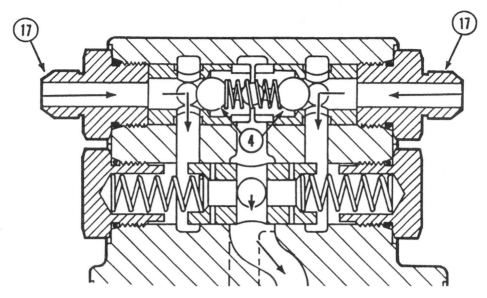

Figure 24.23 Return flow path through the Equa-Trac valve. Courtesy of Case Corporation; Case 7-86.

distribution block has drilled passages open to the inlet and outlet ports of the rotating cylinder block. As pressurized oil enters the piston cavity, it forces the piston outward, moving the follower down the ramp of the cam ring. This motion causes the cylinder block to rotate, turning the wheel hub. As the wheel hub continues to turn, the follower rides up the opposite side of the ramp. This action moves the piston back into its bore, exhausting the oil through the distribution block. The return port of the oil distribution block now directs this oil out of the motor, back to the valve, then back to the main pump.

To provide a smooth power flow, four pistons are in the power stroke, four pistons are in the exhaust stroke, and two are in a neutral stage. Opposing pistons in the cam ring are in the same position of their stroke.

This system requires back pressure on the return side during operation so that the piston remains in contact with the cam ring. This is provided by the charge pressure circuit (low pressure side of the closed circuit) of the hydrostatic system.

Motor Case Flush System

On some models extra lines are added to provide cooled hydraulic oil to the wheel motors. This cooled oil flushes the motor case at all times, removing heat. The motor case flush system consists of a line tap between the oil cooler and the reservoir (Figure 24.26). The tap consists of a tee fitting and a 15-psi check valve that forces part of the cooler return oil to the rear wheel motors before it can go to the reservoir. Cooled oil flow through a 0.5-in. hose to a bulkhead tee fitting where is divided to each motor. The oil flows around the inner components of each motor and then is returned to the reservoir.

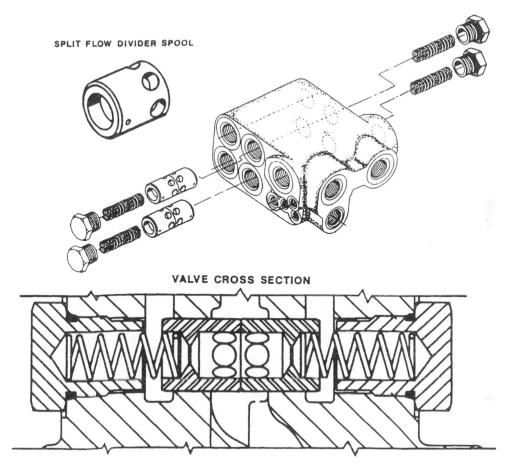

Figure 24.24 Split-flow divider valve assembly. Courtesy of Case Corporation; Case 7-87.

TROUBLESHOOTING THE REAR WHEEL DRIVE SYSTEM

The Equa-Trac valve is a key troubleshooting point. Figure 24.27 shows the points on the valve which are involved in the troubleshooting function.

The first test is for sluggish performance, low charge pressure and/or excessive leakage (oil temperature 150°F). Make certain the outside of the valve is clean before opening any test ports. Install a 10,000-psi gauge in both forward and reverse test ports. Install a 500-psi gauge in both pilot pressure ports (engage and disengage). With the rear wheel drive on, record pressure readings in both forward and reverse. Also check the pressures with the rear drive system off. This will isolate the problem as being in the main system or rear drive system. If the correct pressures can be achieved with the rear drive of, the problem is in the rear drive. If the pressures are incorrect with the unit off, the problem is in the main system.

Sluggish performance can be caused by low charge pressure. This does not allow system pressure to reach normal operating pressure. Causes of low charge pressure could be

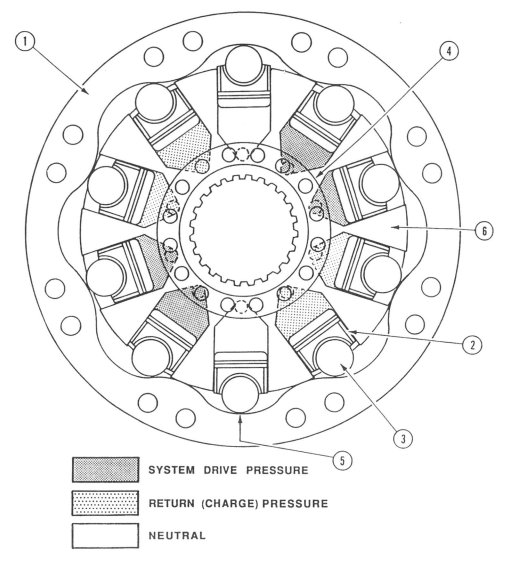

SYSTEM DRIVE PRESSURE

RETURN (CHARGE) PRESSURE

NEUTRAL

1. CAM RING
2. PISTON ASSEMBLY
3. CAM FOLLOWER
4. OIL DISTRIBUTION BLOCK
5. NEUTRAL POSITION
6. CYLINDER BLOCK ASSEMBLY

Figure 24.25   The cam lobe motor. Courtesy of Case Corporation; Case 7-89.

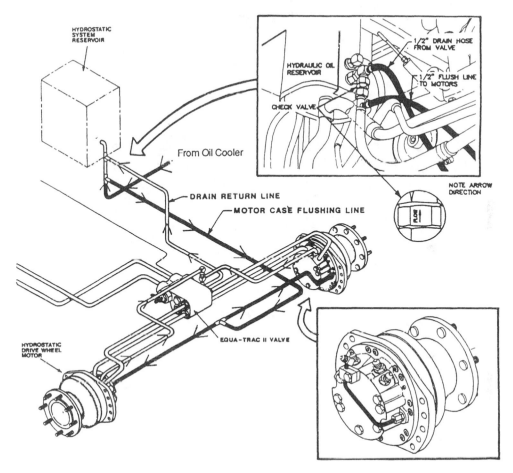

Figure 24.26 Motor case flush system. Courtesy of Case Corporation; Case 7-91.

a plugged filtration system, collapsing suction line, malfunctioning charge or shuttle relief valves, or excessive internal leakage.

Leakage from the wheel motors should not exceed 1.5 gpm with the machine moving at normal system pressure. The Equa-Trac valve leakage should not exceed 1.0 gpm. Motor and valve leakage can be checked by performing the following:

1. Disconnect and plug the motor case flushing hose at the wheel motor tube fitting (if equipped with flush lines).
2. Disconnect case drain hose at the Equa-Trac valve and place it in a clean bucket.
3. Allow oil to escape into the container while operating the machine with the rear wheel drive on. Time how fast the oil leaks out of the system and compare this to specifications. *Important*: Always keep safety and cleanliness in mind.

Maximum drive pressure varies for the type of machine involved. Check the manufacturers specification for the application being serviced. If drive pressure cannot be reached, check the pressure at the foot-n-inch valve with the rear wheel drive off. The

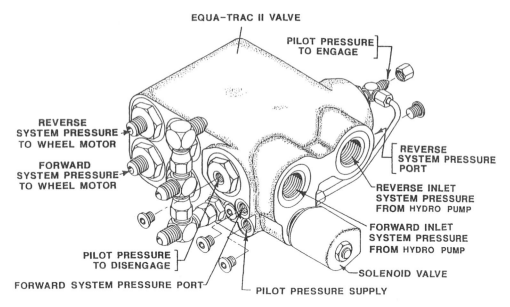

EQUA–TRAC II VALVE

PILOT PRESSURE TO ENGAGE

REVERSE SYSTEM PRESSURE TO WHEEL MOTOR

FORWARD SYSTEM PRESSURE TO WHEEL MOTOR

REVERSE SYSTEM PRESSURE PORT

REVERSE INLET SYSTEM PRESSURE FROM HYDRO PUMP

FORWARD INLET SYSTEM PRESSURE FROM HYDRO PUMP

SOLENOID VALVE

PILOT PRESSURE TO DISENGAGE

FORWARD SYSTEM PRESSURE PORT

PILOT PRESSURE SUPPLY

Figure 24.27 Location for pressure gauge installation on the Equa-Trac valve. Courtesy of Case Corporation; Case 7-93.

problem could be in the main system. Review the troubleshooting guide for the main system.

System not Functioning

Turn the ignition switch to the ON position. With someone located near the rear axle, turn the rear wheel drive switch to the ON position. A distinct click should be heard as the valve spool shifts. If the solenoid valve spool is not moving, perform the following:

Check for battery voltage at the wire that feeds the solenoid valve.
Check for improper grounding of the solenoid. Run a jumper wire from the solenoid post to ground to check this connection.
Check the resistance of the solenoid coil; it should be 7–9 OHMS.

If batter voltage is at the valve and the coil resistance is within specifications, check the valve for debris. If clean and still not functioning, replace the solenoid valve. If battery voltage is not present at the valve, check the following:

Check for battery voltage to the fuse servicing the solenoid.
Check for a blown fuse.
Check for a pinched or broken wire between the switch and the valve.

System not Functioning, Selector Spool not Moving

The oil temperature should be 150°F. Make certain that the outside of the valve is clean before removing any test ports. Install 500-psi pressure gauges in both pilot ports (engage and disengage). Install 10,000-psi gauges in both the forward and reverse drive ports. Record pressures with the machine moving and with system on and off. Check the manufacturer's specifications for the desired pressure level. Typical test pressures are shown in Table 24.2. If pilot pressure readings are not correct, check the solenoid valve for failure (previous test). If pilot pressure readings are correct, but the unit still does not drive, the selector spool is not shifting. Check the 0.020-in. orifice on the spring end of the selector spool; it must be open. If the orifice is open, remove the selector spool and check to see that it moves. Also check for excessive wear. Reinstall the spool and check for free movement. If the Equa-Trac valve has failed (the spool continues to stick), replace the entire valve.

After having completed the installation or servicing of the rear wheel drive system, the manufacturer's start-up procedure must be performed to insure adequate bleeding and flushing of the newly installed or serviced hydraulic components. The instructions must be carried out as specified. Any alteration of the process will defeat its purpose, which is to bleed air out of the system and flush any possible contamination from the closed loop. Special attention must be given to the 0.5-in. limited movement of the hydrostatic control lever. It is critical not to allow excessive oil flow during this procedure. If at any time while performing these functions a malfunction occurs that causes the system to be reopened, it is necessary to start from the first recommended step.

TROUBLESHOOTING THE TOTAL CIRCUIT

The common component within the total circuit is the hydraulic reservoir. Traction drive components function as an independent circuit responding to the control of the hydrostatic pump and the foot-n-inch valve.

The remaining hydraulic force and motion functions are supplied with fluid from the auxiliary double-gear pump or single-gear pumps as appropriate to the circuit requirements associated with the various auxiliary systems.

Typical components employed in a total combine hydraulic system are shown in the schematic of Figure 24.28. Seven functions can be controlled by electrically actuated valving in a manifolded assembly to minimize piping and physical space on the machine. Solenoid actuation simplifies the control functions.

Figure 24.29 illustrates a typical combine electro-hydraulic system by means of sectional views of the components. The hydraulic schematic of Figure 24.28 using ISO symbols indicates by a centerline enveloping the symbols that the individual valves are in a manifolded stack assembly. The valves in this manifolded assembly consist of the control for the feeder clutch cylinder, reel lift, unloader swing cylinder, reel linear movement cylinders, header lift cylinders, and other associated valving. The reel drive control is shown as a separate entity as is the power steering componentry. The diagram of Figure 24.29 may assist in visualizing the functions of the various controls.

Pressurized Fluid for the Auxiliary Functions

The gear-type hydraulic pumps are located directly below the reservoir. The tandem pump of Figure 24.30 is located directly below the reservoir. The larger main pump in this tandem assembly contains a flow divider and steering system relief valve as shown in Figure 24.31 in symbolic format. The gear pumps are driven by a double V-belt and pulley system derived directly from engine power. Belt tension is maintained by a spring and rod with a spacer. An attenuator is installed in the supply line to the steering system to smooth out the

Table 24.2 Typical Test Pressures for Rear Drive Circuits (Case 7-97)

Operation mode	Forward port	Reverse port	Forward line to wheel motor	Reverse line to wheel motor	Pilot pressure to engage	Pilot pressure to disengage
Rear wheel drive ON; Forward	System pressure* 1000–5600 psi	Charge pressure 160–190 psi	System pressure* 1000–5600 psi	Charge pressure 160–190 psi	Charge pressure 160–190 psi	Case pressure 20–40 psi
Rear wheel drive ON; Neutral	Charge pressure 220–240 psi	Charge pressure 220–240 psi	Charge pressure 220–240 psi	Charge pressure 220–240 psi	Charge pressure 220–240 psi	Case pressure 20–40 psi
Rear wheel drive ON; Reverse	Charge pressure 160–190 psi	System pressure* 1000–5600 psi	Charge pressure 160–190 psi	System pressure* 1000–5600 psi	Charge pressure 160–190 psi	Case pressure 20–40 psi
Rear wheel drive OFF; Forward	System pressure* 1000–5600 psi	Charge pressure 220–240 psi	Case pressure 20–40 psi	Case pressure 20–40 psi	Case pressure 20–40 psi	Charge pressure 160–190 psi
Rear wheel drive OFF; Neutral	Charge pressure 220–240psi	Charge pressure 220–240 psi	Case pressure 20–40 psi	Case pressure 20–40 psi	Case pressure 20–40 psi	Charge pressure 220–240 psi
Rear wheel drive OFF; Reverse	Charge pressure 160–190 psi	System pressure* 1000–5600 psi	Case pressure 20–40 psi	Case pressure 20–40 psi	Case pressure 20–40 psi	Charge pressure 160–190 psi

*Maximum system (drive) pressure specifications vary on the type of combine (rice, corn, or grain) being serviced. Check for the proper pressure on the specifications page.

Source: Case Corporation.

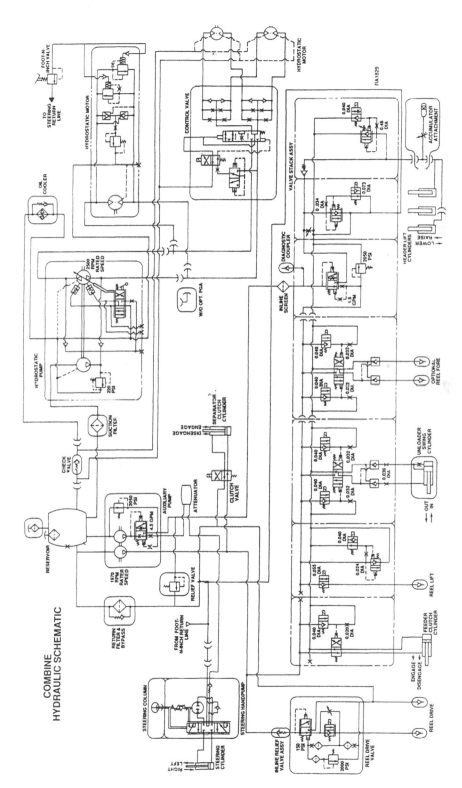

Figure 24.28 Hydraulic schematic for the total combine force and motion functions employing ISO symbols. Courtesy of Case Corporation.

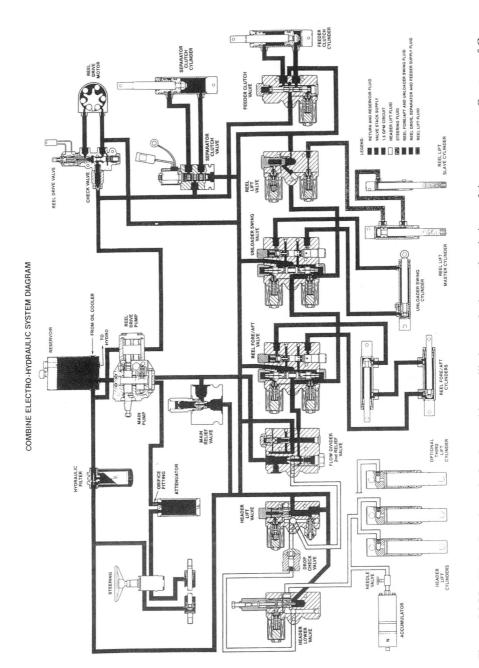

Figure 24.29 Hydraulic schematic for the combine auxiliaries employing sectional views of the components. Courtesy of Case Corporation.

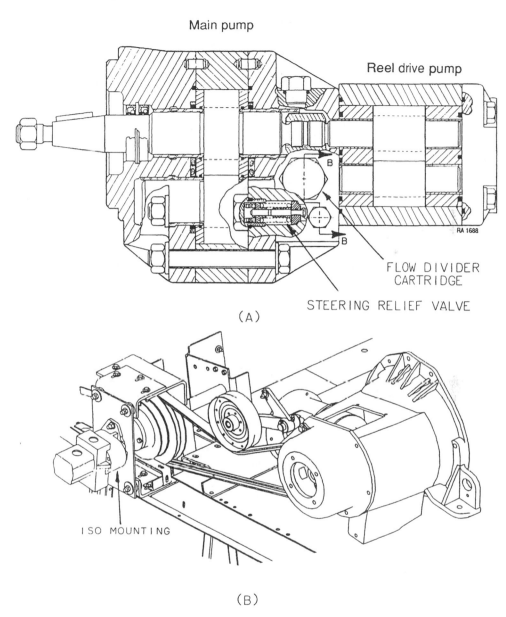

Figure 24.30 Pump assembly for auxiliaries. (A) Main and reel drive pump with flow divider shown in sectional view. (B) Drive assembly. Courtesy of Case Corporation; Case 3-12.

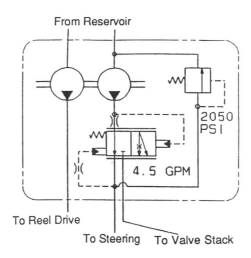

Figure 24.31 Flow divider for power steering supply. The flow divider is a cartridge structure assembled into the main pump housing. Courtesy of Case Corporation; Case 3-19.

supply pressure (Figure 24.32). Figure 24.33 illustrates the power steering assembly. The metering pump can function as a power driven pump with the steering wheel as the input power in the event of engine failure for emergency purposes; however, the steering effort will be greater. The metering pump when powered acts to cancel the turning signal as the cylinder delivers power to the steering functions.

In-Line Filter

After the main pump flow has been exposed to the primary relief valve, it flows to the in-line nonbypass high pressure filter. This filter is located on the flow divider section of the main valve stack. It is used to prevent large foreign material from entering the valve stack (Figure 24.34).

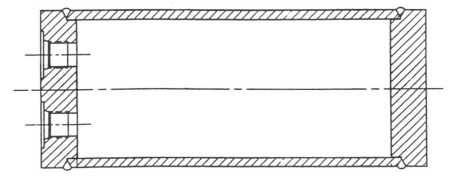

Figure 24.32 The attenuator is a canister that holds a trapped volume of air and oil. It is used to remove pulsations in the oil flow that can cause noise in the steering hand pump. The inlet to the attenuator is a special orifice elbow fitting. Courtesy of Case Corporation; Case 3-23.

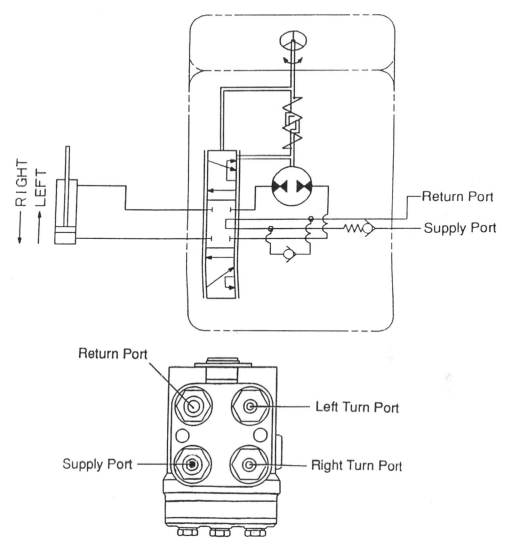

Figure 24.33 Power steering assembly in neutral position. Courtesy of Case Corporation; Case 3-25.

Main Stack Valve

The main valve stack is located at the outside left rear corner of the cab. It is an open center system with seven main sections that control the following functions:

Secondary Flow Divider and Relief Valve

This valve divides oil flow between the secondary circuits (reel lift, reel fore and aft, unloading auger swing) and the third priority circuit (i.e., the header lift). It also contains the relief valve for the secondary circuits and a pressure test port.

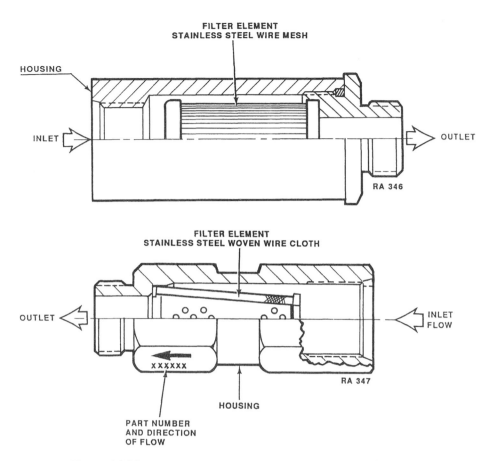

Figure 24.34 In-line filter. Courtesy of Case Corporation; Case 3-36.

Reel Lift and Lower

This circuit utilizes a master/slave cylinder arrangement on the header. When not using this function, open center flow for the secondary circuits is directed to the reservoir through the reel lift valve.

Reel Fore and Aft

This circuit utilizes two double-acting cylinders on the header. When not using this function, open center flow is directed to the reservoir through the reel lift valve.

Unloader Swing

This circuit utilizes one double-acting cylinder on the unloader tube. When not using this function, open center flow is directed to the reservoir through the reel lift valve.

Header Lift and Lower Valves

This circuit utilizes two single-acting ram-type cylinders. A third cylinder is optional and an accumulator is standard. When the header lift circuit is not being used, its fluid goes to the reservoir through the header lift valve.

Feeder Clutch Control Valve

This valve controls a double-acting cylinder that engages and disengages the feeder. Oil supply is from the reel drive circuit and return is through the main valve stack.

Main Valve Stack Solenoid Functions

It is very important to be able to separate the hydraulic and electrical portions of an electro-hydraulic system if problems exist. The easiest way to separate the two is by verifying that the correct solenoids are energized at the proper times. This can be checked with a volt ohm meter or by checking for magnetism at the solenoid. If the electrical tests correctly and the system still does not work, the problem is in the mechanical-hydraulic portion of the system. A list of which solenoids must be energized to operate the given hydraulic force and motion functions is provided (Figure 24.35).

There are two types of electric solenoids in use: Orange—Constant duty, "O" imprinted on coil body, and Yellow—intermittent duty, "Y" imprinted on coil body. Solenoids 1, 2 and 9 of Figure 24.35 are constant duty (13.5 ohms resistance); 3–8 are intermittent duty (3.5 ohms resistance).

Secondary Flow Divider

All oil not used for steering, flowing from the main pump, enters the secondary flow divider valve after passing through the in-line pressure filter. This is the third valve from the front of the seven in the stack. When viewing the valve as it sits on the combine, the secondary flow divider is located behind the plug on the left side of the third valve. The inlet for the valve is directly across (right side). *Note*: The secondary flow divider must be removed from the left side of the valve.

The secondary flow divider spool has an orifice in the inlet end and a hollow bore with four spill ports midway in the spool body (Figure 24.36). There is also a spring and orifice plate on the backside of the flow divider. The spill ports line up with a groove and cored passage to the secondary relief valve and 2.0 gpm circuits.

In operation the pump flow enters the secondary flow divider spool cavity at the pump port. The orifice in the end of the spool allows the proper amount of fluid through to the 2.0 gpm circuits (reel lift, auger swing, and reel fore-and-aft linear movement). The remaining pump flow builds up pressure against the spool, shifting it against its spring to open the cored passage to the header lift section. This flow divider provides second priority to the 2.0 gpm circuits, and third priority to the header lift. The orifice plate in the end of the flow divider dampens the movements of the unit.

Pilot Operated Secondary Relief Valve

The reel lift, reel fore-and-aft, and unloading auger swing systems are supplied with 2.0 gpm from the secondary flow divider. To insure uninterrupted header lift flow, a secondary relief valve is incorporated into the same valve block as the secondary flow divider. The relief valve components are accessible through a jam nut and plug assembly directly below the pump inlet port. This is an adjustable pilot operated relief valve (Figure 24.37).

As system pressure approaches 2050 psi, the relief valve starts to open. Since this is a pilot operated relief valve, the pilot poppet must move first. The increased system pressure pushes the pilot poppet against its spring; this allows the oil trapped in the main poppet cavity to escape to reservoir. This is pilot flow, not main relief flow. Main relief flow occurs

1 - Header Lower   7 and 8 - Reel Raise
2 - Header Raise   7 and 4 - Reel Forward
8 - Reel Lower   7 and 3 - Reel Aft
9 - Feeder Engage   7 and 5 - Unloader Extend
                                    7 and 6 - Unloader Retract

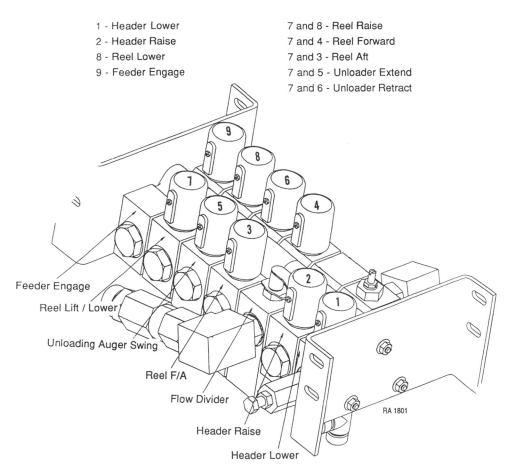

Figure 24.35   Solenoid coil identification and solenoids to be actuated to provide the desired force and motion function. Courtesy of Case Corporation; Case 3-39.

when the increased pressure causes the main poppet to rise directing the major flow to the reservoir.

Reel Lift Valve

Oil flowing through the 2.0 gpm orifice of the secondary flow divider proceeds through an open center flow port. This port is common to the following valves: reel fore-and-aft, unloading auger swing, and reel lift. The valves are connected by small jumper tubes between each section. When disassembled, two flow ports can be seen between the valves: one is supply, the other return. The reel lift valve is the controller of the 2.0 gpm of open center flow. This valve must function before any other secondary circuit will operate.

The reel lift valve is the sixth section of the valve stack. It contains two solenoids that

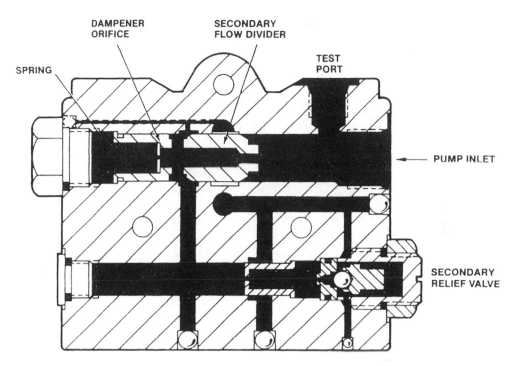

Figure 24.36 Valve stack-flow divider section. Courtesy of Case Corporation; Case 3-42.

control one normally open poppet, and one normally closed poppet. There is also one secondary poppet (Figure 24.38).

Reel Lift Valve in Neutral

Approximately 2.0 gpm flow from the secondary flow divider enters the valve block and is directed to the normally closed primary poppet where it is blocked. Fluid also is directed through the primary orifice to the normally open primary poppet and back to the reservoir. Since all the flow cannot pass the orifice, pressure increases and pushes the secondary poppet over, thereby opening the passage to the reservoir.

When the valve is in the neutral position, the weight of the reel will try to force fluid out of the lift cylinders and back to the cylinder port of the valve. Since the fluid cannot pass the normally closed primary poppet, until a pressure of about 1400 psi is attained, the reel will normally be held in place.

Reel Lift Valve in Raise

To raise the reel, the reel lift switch is moved to the rear energizing both the reel solenoid (8, Figure 24.38) and jammer solenoid (7). Approximately 2.0 gpm minimum flow from the secondary flow divider enters the valve block and is directed through the primary orifice to the now closed primary poppet and to the back side of the secondary poppet. Since the primary poppet is closed, oil pressure and spring pressure will push the secondary poppet closed. This blocks the open center flow, causing the system to go "on demand." Fluid will

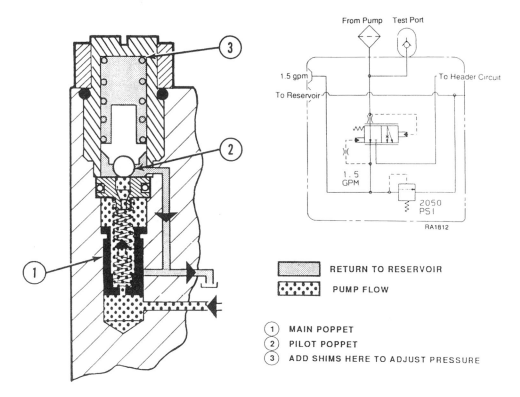

Figure 24.37 Pilot operated secondary relief valve. Courtesy of Case Corporation; Case 3-44.

then be directed to the now open reel primary poppet (solenoid 8) and on to the reel lift cylinders to raise the reel.

The reel lift cylinders are a master and slave arrangement. The flow to the slave cylinder comes from fluid displaced by the piston of the master cylinder. The slave cylinder has the same area on the base end as the master cylinder has on the rod end. When the master cylinder is moved, fluid is forced to the slave cylinder. This eliminates uneven lift of the reel. The slave cylinder is a ram-type cylinder while the master cylinder is a piston-type.

The cylinders will rephase with each other by momentarily holding the lift control valve "on demand" at the top or bottom end of the cylinder stroke. The master cylinder has rephasing orifices for this purpose. This will compensate for some normal seal leakage. All air must be purged from the slave cylinder for this to work. There is an air bleed screw on the top of the slave cylinder for this purpose. When the valve is in the lower position, the weight of the reel will force fluid out of the lift cylinders and back to the cylinder port of the valve. Since the primary poppet at this port is now open, the fluid from the cylinders will join with the pump flow and return to the reservoir. Reel "lower" has electrical priority over other secondary functions. This prevents raise when another function is selected.

Unloading Auger Swing Valve

The unloading auger swing valve is the fifth valve of the stack (Figure 24.39). It contains two solenoids with the primary poppets, a directional spool, two plastic load check valves,

To lower the reel, the operator moves the switch forward, energizing the reel solenoid (8).

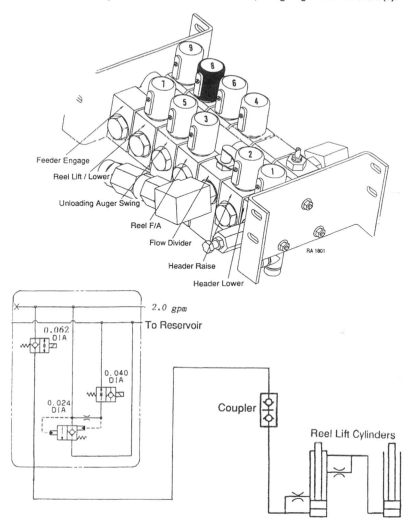

Figure 24.38 Reel lift control circuit. Courtesy of Case Corporation; Case 3-52.

and two controlling orifices. Approximately 2.0 gpm (same as reel lift) is available from the secondary flow divider. This flow enters the swing valve and is directed to the primary poppets which block off the flow. The flow is also blocked by the spring centered main spool. When in neutral, this pump flow can return to the reservoir through the reel lift valve. The unloading auger is held in position by two check valves within the valve block. These trap fluid on both sides of the cylinder.

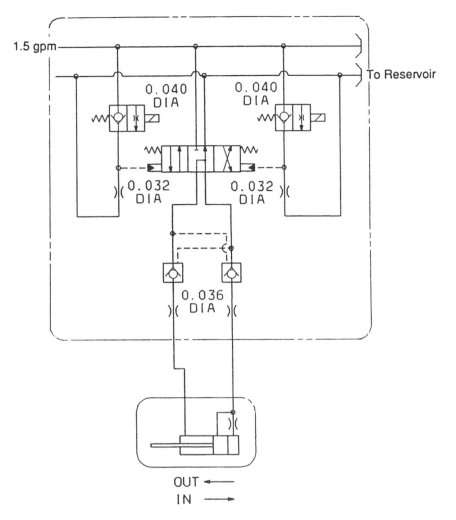

Figure 24.39 Unloading auger swing valve in the retract position. Flow through the reel valve is the same as the flow described in the extend position (see text). Flow through the auger swing valve is similar, but in the opposite direction. The fluid from the valve is directed to the double-acting cylinder that moves the unloader tube to any desired position within its range. Note: If the unloading auger swing valve is ever disassembled or replaced, verify that the orifices in the work ports (1301637C1, .037 in.) are in place. If not, they must be removed from the work ports of the old valve and installed in the new valve. Failure to install these orifices will result in unrestricted flow to the auger swing cylinder, allowing the auger to swing rapidly and possibly causing damage to the auger saddle and auger pivot elbow. Courtesy of Case Corporation; Case 3-59.

Unloading Auger Extend

A switch located on the steering post is used to energize the solenoids on the swing valve and the jammer solenoid on the reel lift valve. When the switch is moved into the extend position, it energizes the jammer solenoid 7 and the auger extend solenoid 5. When the number 7 solenoid is energized, the primary poppet closes, blocking flow to the reservoir.

Pressure then builds behind the secondary poppet. The oil and spring pressure push the poppet against it seat, closing the passage to the reservoir. This blockage of the fluid going to reservoir now puts 2.0 gpm on demand for the auger swing valve (Figure 24.39).

Pressurized fluid from the reel lift valve enters the unloader swing valve and is directed to both normally closed primary poppets. The closed poppet blocks fluid flow while the open poppet (solenoid energized) allows fluid to pass, directing it to the main spool. The main spool moves against a spring opening a flow path between the main spool and the inlet passage. The fluid is directed to a check valve and piston. This pressurized fluid opens the check valve, opening the passage to the cylinder. The piston is pushed toward the opposite check valve, unseating it, opening a passage to the reservoir for the return oil from the piston.

The unloader swing electrical circuit contains limit switches that allow the unloader swing switch to be located in either detent position without causing damage to the hydraulic system. When the switch on the steering post is moved to the retract position, it energizes the jammer solenoid 7 and the auger retract solenoid 6. Flow through the reel valve is the same as the flow described in the extend position. Flow through the auger swing valve is similar, but in the opposite direction.

The fluid from the valve is directed to the double-acting cylinder that moves the unloader tube to any desired position within its range. *Note*: If the unloader auger swing valve is ever disassembled or replaced, verify that the orifices in the work ports are in place. If not, they must be removed from the work ports of the old valve and installed in the new valve. Failure to install these orifices will result in unrestricted flow to the auger swing cylinder allowing the auger to swing rapidly, possibly causing damage to the auger saddle and auger pivot elbow.

Unloader Swing Cylinder Decelerator

To prevent damage to the unloader saddle and/or tube, a decelerator is designed into the unloader swing cylinder. The piston end of the cylinder has two orifices of different sizes drilled through the cylinder barrel. These orifices are 0.020 and 0.063 in. in diameter and are exposed to a common channel. The piston itself has two sealing rings that are spaced apart the same distance as the orifices (Figure 24.40).

In the retract position, the cylinder returns at its normal rate of speed. The fluid returns to the valve through the 0.020 inch and 0.063 inch orifices. This continues until the first piston seal ring covers up the 0.063-in. orifice. The remaining travel (1–2 ft) to the saddle has the oil returning only through the 0.020-in. orifice which causes the deceleration. Initially the auger will leave the saddle slowly because of oil only flowing through the 0.020-in. orifice (the 0.063-in. orifice is blocked by the piston seal ring). As soon as the cylinder extends the length of the piston, the 0.063-in. orifice is uncovered. The remaining travel outward will be at a normal rate of speed with flow through both orifices. Once both are open, the speed at which the auger moves is controlled by the orifices (0.037 in.) in the work ports of the unloader swing valve (Figure 24.40).

Reel Fore-and-Aft Valve

The reel fore-and-aft is the fourth valve of the stack (Figure 24.41). It contains two solenoids with normally closed primary poppets, a directional spool, and two plastic load check valves. It is optional for all models.

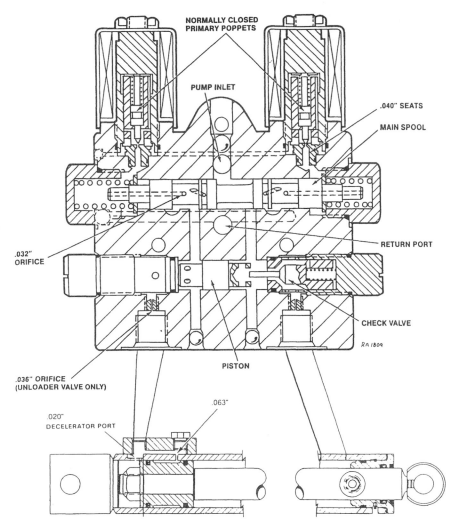

Figure 24.40 Unloader swing cylinder decelerator. To prevent damage to the unloader saddle and/ or tube, a decelerator is designed into the unloader swing cylinder. The piston end of the cylinder has two orifices of different sizes drilled through the cylinder barrel. These orifices are .020 in. and .063 in. in diameter and are exposed to a common channel. The piston itself has two sealing rings spaced apart the same distance as the orifices are. Courtesy of Case Corporation; Case 3-60.

In the neutral position no solenoids are energized, the main spool centers, both primary poppets are closed, and the pump flow goes to the reservoir through the reel lift valve. The fluid in the cylinders is checked by the two plastic load checks. The fluid in the reel fore-and-aft cylinders and springs, causes the check valves to seat. The tapered end of each check valve is exposed to reservoir when the main spool is centered. This is to aid in seating the check valves. Depressing the switch on the right-hand console forward energizes solenoids number 7 and 4.

The reel fore-and-aft valve is the same valve as the previously discussed unloader

Reel Forward Position

Depressing the switch on the right-hand console forward will energize solenoids number 7 and 4.

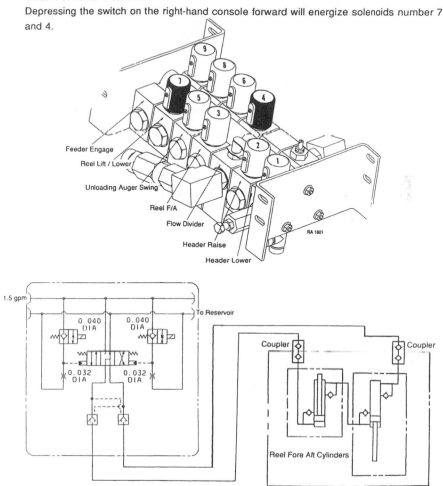

Figure 24.41 Reel fore and aft circuit in reel forward position. Courtesy of Case Corporation; Case 3-63.

swing valve. It also receives its fluid from the secondary flow divider. Anytime the reel fore-and-aft switch is depressed, solenoid number 7 of the reel lift valve is energized, thus putting the 2.0 gpm on demand, in addition to either of the solenoids of the reel fore-and-aft valve. This provides fluid to the reel fore-and-aft valve that would otherwise go to the reservoir through the reel lift valve.

Pressurized fluid from the reel fore-and-aft valve is directed to the piston end of either fore-and-aft cylinder. Both ports of each cylinder contain rephasing check valves. If the cylinders get out of phase, holding the reel fore-and-aft switch on demand in either direction synchronizes the cylinders. Depressing the switch on the right-hand console energizes solenoids 7 and 3 for reversal.

Header Lift Valve

The header lift valve is the second valve of the stack. Pump flow to the header lift valve is third priority (Figure 24.42). This is established by the secondary flow divider. Once the 2.0 gpm circuits are satisfied, the pressure build-up on the face of the secondary flow divider causes it to shift against the spring. This opens a passage to the header lift valve. Flow to this valve is approximately 15 gpm minimum.

Pump flow enters the valve block from the secondary flow divider and is directed past the variable raise rate valve. This valve can be used to slow down the speed of header raise by allowing oil a direct path to return (as opposed to going to the lift cylinders). Fluid not going past the variable raise rate valve will proceed to a 0.034-in. orifice in the face of the secondary raise poppet. A small amount of fluid passes this orifice, reduces in pressure, and proceeds to the normally open primary poppet seat. This seat is 0.073 inch in diameter, and allows the fluid to pass on to the reservoir since the poppet is open. Since all the pump flow cannot pass the 0.034-in. orifice in the face of the secondary raise poppet, pressure increases and pushes the secondary poppet against its spring and hence off its seat. This movement allows all remaining header flow to pass to the reservoir, and establishes a third priority open center flow.

Header Lower Valve

The header lower valve is the first valve in the stack. Its purpose is to provide a controlled lowering function for the header and to hold the header stable. When a header is attached to the combine and raised off the ground, the weight of the header tries to force the oil out of the lift cylinders. The pressure created by the header weight backs up into the header lower valve. A small amount of oil will pass through a 0.046-in. orifice in the face of the secondary lowering poppet. It proceeds to the back side of the secondary lowering poppet, and then to the normally closed primary poppet and seat. The primary poppet seat is 0.040 in. in diameter and is blocked off at this time. Because there is no oil flow, oil pressure equalizes on both sides of the secondary lowering poppet. The spring on the backside of the secondary lowering poppet, along with greater surface area than the front, seats the secondary lowering poppet. The oil pressure from the lift cylinders also seats the drop check valve. This poppet is located in the header lower valve port (Figure 24.43).

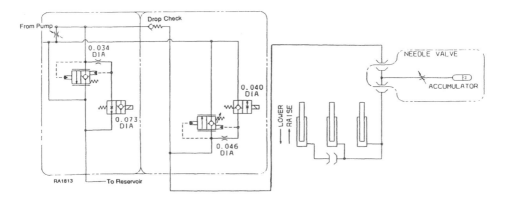

Figure 24.42 Header lift valve circuit. Courtesy of Case Corporation; Case 3-67.

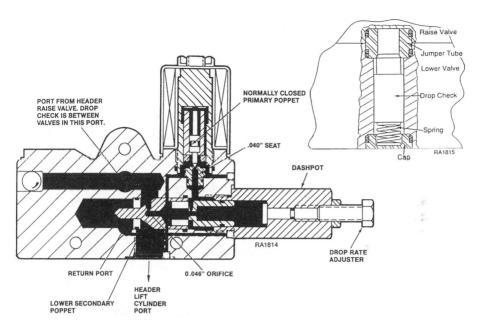

Figure 24.43 Header lower valve and drop check. Courtesy of Case Corporation; Case 3-68.

Header Raise

A rocker switch located in the hydrostatic drive propulsion control handle is used to energize the solenoids of the header valve to raise and lower the header. When the switch is pressed into the raise position, it energizes the raise circuit solenoid (2, Figure 24.43) of the valve which contains a normally open primary poppet, causing the poppet to close.

Pump flow enters the header raise valve block through a jumper tube from the secondary flow divider valve. It is directed through a 0.034-in. orifice in the face of the secondary raise poppet to the back side of the raise secondary poppet. From here, oil will proceed to the now closed 0.073-in. primary poppet seat. Since the primary poppet is closed, pressure builds between the back of the secondary raise poppet and the closed primary seat. Oil pressure and spring pressure push the raise secondary poppet against its seat. The seating of the secondary raise poppet causes the header lift circuit to go on demand.

Because header oil flow can no longer go to the reservoir, it proceeds to the lift check in the header lowering valve. Pump pressure must build to a level equal to the sum of the pressure in the lift cylinders and the pressure of the lift check valve spring. This prevents the header from dropping slightly before it begins to raise.

When the lift check valve is pushed off its seat, the fluid is directed to the front side of the lower secondary poppet, through the 0.046-in. orifice. (*Note*: the header lower valve is in the neutral state during header raise.) Oil proceeds to the back side of the lower secondary poppet and on to the normally closed primary poppet and seat. Since the lower primary poppet is closed, pressure builds and holds the lower secondary poppet closed. The fluid on the front side of the secondary lowering poppet is now directed to the cylinders to raise the header.

The amount of fluid going to the lift cylinders can be adjusted by turning the variable raise rate valve. Opening the valve allows part of the fluid to go to reservoir instead of the lift cylinders. The farther the variable raise valve is open, the slower the header will raise. This valve gives an adjustability of flow from about 10 to 15 gpm minimum.

Lowering the Header

Pump flow is not required to lower the header; therefore, the header raise valve will be in neutral allowing third priority open center flow to go to the reservoir.

When the rocker switch on the hydrostatic drive propulsion control handle is pressed into the lower position, it energizes the lower circuit solenoid (1) of the header lift valve. When solenoid number 1 is activated, it causes the normally closed primary poppet to open its 0.040-in. seat. This releases the oil that was trapped behind the secondary lowering poppet. The oil pressure caused by the weight of the header forces fluid out of the lift cylinders and back to the cylinder port of the valve. Because the primary poppet has opened, a pressure differential exists between the frontside and backside of the secondary lowering poppet. The pressure on the frontside of the secondary lowering poppet pushes the poppet against its spring, opening a port to reservoir for the lift cylinder fluid (Figure 24.44).

The speed at which the header drops is controlled by how large the header is and by how far the secondary lowering poppet moves off its seat. The movement of the secondary lowering poppet can be controlled by the drop rate adjusting bolt in the dashpot housing. This adjustment overrides the effects of different size headers.

By turning the bolt out of the housing, the poppet moves farther from its seat, exposing a larger port to the reservoir, and increasing the lowering speed. By turning the bolt into the housing, the poppet stays closer to its seat, creating a smaller port to the reservoir, and decreasing the speed of lowering.

To allow a smooth and controlled opening of the secondary poppet, a dashpot piston is used. As the secondary lowering poppet moves to the right, trapped oil on the inside of the dashpot piston must be displaced through a small orifice. This dashpot piston allows the secondary poppet to open smoothly, which allows the header to be lowered smoothly. *Note*: To increase the speed at which the feeder lowers without a header attached, first raise the feeder and hold on demand for a moment. This charges the cylinder lines with pressure so that when the feeder is to be lowered, the stored pressure pushes the secondary lowering poppet farther off its seat.

Hydraulic System Dampener

The accumulator is used as a hydraulic system dampener for the header lift/lower circuit. The accumulator system consists of an external needle valve, an internal piston within a cylinder, and a Schrader valve (Figure 24.45). It is located adjacent to the left-hand tire and is piped into the header lift/lower circuit.

The gas side of the accumulator piston contains compressed nitrogen while the fluid side is exposed to the header circuit hydraulic fluid (with needle valve open). As the hydraulic oil enters the accumulator, the piston is pressed against the nitrogen gas, absorbing any shock loads. The needle valve can be used to control the speed at which the fluid enters the accumulator. The accumulator can be used during all field and road conditions. When used with autoheader height control, it dampens the reaction of the system depending

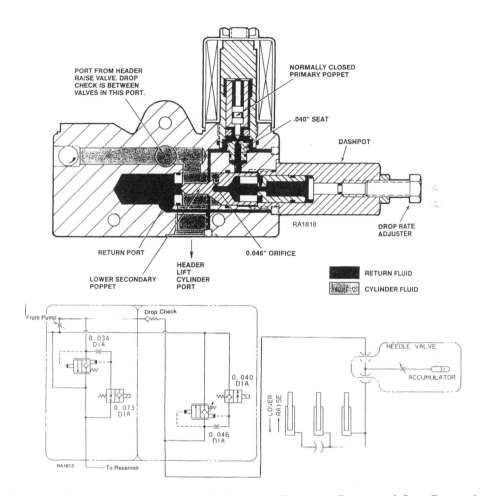

Figure 24.44 Header lower valve with drop rate adjustment. Courtesy of Case Corporation; Case 3-72.

on how far the accumulator needle valve is opened. It is extremely important that the accumulator be used properly for operator comfort.

Adjusting Accumulator Pressure

The pressure in the nitrogen gas side of the accumulator should be adjusted to the weight of the heaviest header that will be used on the combine. To do so:

1. Open the needle valve on the accumulator, start the engine, and lower the header to the ground. Continue to hold the header lower switch for 5 seconds once the header is fully lowered.
2. Close the accumulator valve.
3. Raise the header off the ground 12–18 in. and shut off the engine.

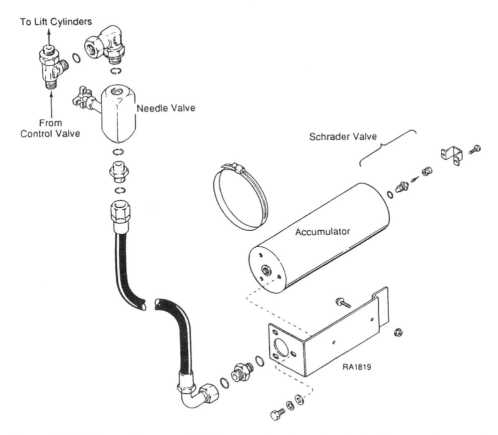

Figure 24.45 Accumulator assembly. The accumulator is used as a hydraulic system dampener for the header lift/lower circuit. The accumulator system consists of an external needle valve, an internal piston within a cylinder, and a Schrader valve. It is located adjacent to the left-hand tire and is plumbed into the header lift/lower circuit. The gas side of the accumulator piston contains compressed nitrogen while the fluid side is exposed to the header circuit hydraulic fluid (with needle valve open). As the hydraulic oil enters the accumulator, the piston is pressed against the nitrogen gas and absorbs any shock loads. The needle valve is used to control the speed at which the fluid enters the accumulator. The accumulator is used during all field and road conditions. When used with the autoheader height control, it dampens the reaction of the system (depending on how far the accumulator needle valve is opened). It is *extremely* important that the accumulator be used properly for operator comfort. It is standard equipment as of 1992. Courtesy of Case Corporation; Case 3-76.

3. Measure the height of the header frame above the ground. (Back side of header).
4. Open the accumulator valve. If the gas pressure is correct, the header will lower a distance of 3–5 in. on a machine with 2 lift cylinders, and 2–3 in. on a machine with 3 lift cylinders.

If the header drops less than the minimum amount, remove the valve guard and release a small amount of gas using a tool to press open the valve. Repeat this procedure starting at step 1. *Note*: Repeat this process until the drop distance is correct, being careful to release the gas in small increments.

If the header drops more than the recommended amount, the accumulator needs recharging. This requires a tank of compressed nitrogen gas and a recharging kit. *Caution*:

Use extreme caution when handling the accumulator: do not expose to extreme heat, do not drop, use only nitrogen gas for recharging, and always use a regulator on the nitrogen supply tank to assure the accumulator does not get exposed to full supply pressure.

The factor precharge on this accumulator is 800 psi.

Feeder Clutch Valve

The feeder clutch valve is the last valve of the stack. It receives fluid from the reel drive pump and returns it to the main stack reservoir port. The reel drive pump is the smaller section of the tandem gear pump.

When the toggle switch is in the OFF position, the solenoid is not energized and the primary poppet is seated (normally closed). Pump flow is supplied to the valve directly from the reel drive pump outlet line. The flow enters the inlet port of the valve and is directed to the primary poppet which blocks the passage. The main spool now is held in the disengaged position by the spring. The oil flow is directed to the main spool which channels oil out to the cylinder rod end. As the cylinder retracts, the fluid displaced from the base end of the cylinder is directed back to the valve, around the spool, and back to the reservoir. Once the cylinder is completely retracted, the system pressure holds the cylinder. The system pressure will be a minimum of 250 psi (because of a 250 psi inlet check valve in the reel drive control valve). When the reel drive is operating, the pressure in the feeder clutch cylinder will be at reel drive pump pressure (Figure 24.46).

When the toggle switch is in the ON position, the valve solenoid is energized causing the main spool to move against the bias spring and reverse flow to the cylinder. Flow is now directed by the spool to the base end of the cylinder. Oil from the rod end is ported to the reservoir through the valve.

Separator Clutch Valve

The reel drive system obtains its oil from the small section of the tandem pump. It is rated for 10 gpm at 500 psi when tested at the couplers. Before the reel drive itself receives fluid, two other functions are connected into this system: the feeder clutch engage valve and the separator clutch engage valve.

The separator engagement control valve is a four-port, solenoid operated, hydraulic directional control valve (Figure 24.47). The separator control valve is located above the PTO drive housing at the front left side of the engine.

A spring located on top of the spool pushes the spool downwards. Fluid from Port B enters the holes in the center of the valve and is directed past spool lands to Port D. Return fluid from port A is directed to the reservoir through the inside of spool to Port C. When the solenoid is energized the spool is attracted to the magnetic field. With the spool moved up, fluid from Port B enters the holes in the center of the valve and is directed past the spool lands to Port A. Return fluid from Port D is directed to the reservoir through the inside of the spool to Port C. *Note*: Once the cylinder reaches the end of its stroke, the flow through this circuit stops. The circuit continues to feel reel system pressure. The control valve has no neutral position.

Reel Drive Valve

The reel drive valve is located on the left side of the combine, just below the valve stack. The valve receives the entire output of the reel drive pump once the feeder engage and

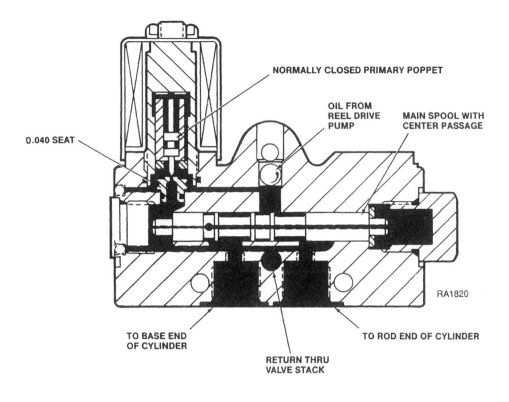

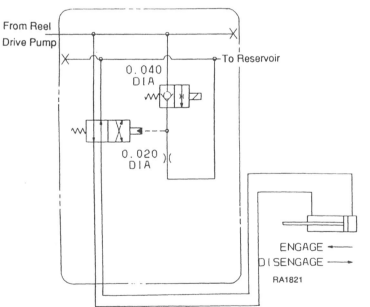

Figure 24.46   Feeder clutch valve and circuit. Courtesy of Case Corporation; Case 3-80.

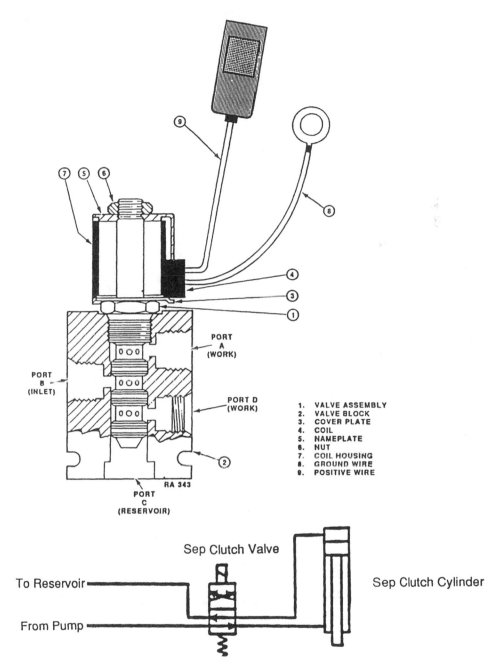

1. VALVE ASSEMBLY
2. VALVE BLOCK
3. COVER PLATE
4. COIL
5. NAMEPLATE
6. NUT
7. COIL HOUSING
8. GROUND WIRE
9. POSITIVE WIRE

PORT A (WORK)

PORT B (INLET)

PORT D (WORK)

PORT C (RESERVOIR)

RA 343

Sep Clutch Valve

Sep Clutch Cylinder

To Reservoir

From Pump

Figure 24.47 Separator clutch valve. Courtesy of Case Corporation; Case 3-84.

separator engage valves are satisfied. The valve then allows a certain amount of flow (selected by the operator) to go to the reel drive motor, with the remaining oil directed to the reservoir.

Two sectional views of the reel drive valve are shown in Figure 24.48. The operator selects the speed of the reel by rotating the reel speed knob. This sends current to the reel drive valve solenoid. Current only flows to the solenoid when the feeder is engaged. The strength of the magnetic field in the solenoid determines the amount of pintle movement. Initially, the main spool is blocking off the reservoir port because the bias spring pushes the main spool to the left. Full pump volume (10 gpm at full throttle) enters the valve and is

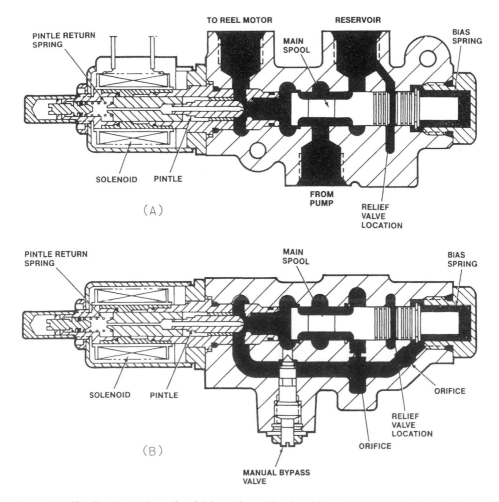

Figure 24.48 Sectional view of reel drive valve with solenoid energized. (A) Section showing reel motor, reservoir and pump inlet porting. (B) Section showing manual bypass valve. Courtesy of Case Corporation; Case 3-88 modified.

exposed to the main spool. Some of the flow is directed by the main spool, past the pintle, and out of the valve to the reel drive motor. A small amount of this fluid is directed through the signal channel, around the manual bypass, and to the right-hand end of the main spool. Once the volume to the reel drive motor is satisfied, pressure increases (at the inlet) an additional 100 psi to move the main spool to the right. This directs the oil not used by the reel drive motor to the reservoir. The bias spring on the end of the main spool assures priority flow to the reel drive motor. The reel drive motor is the path of least resistance because of the bias spring. The 0.031-in. orifice in the signal passage is used to bleed off normal internal valve leakage so that the reel does not turn when the valve is shut off. It also aids in relieving line pressure so the couplers disengage better.

The reel drive valve itself has a relief function built into it. If the signal pressure on the right-hand end of the spool exceeds 2000 psi, a spring-loaded poppet will move off its seat, exposing the signal cavity and oil around the bias spring to the reservoir. This drains the fluid from the right-hand end of the spool allowing the incoming fluid (from the pump) to force the main spool to the right. This makes the reservoir port of the path of least resistance so most pump volume will go there. The 0.025-in. orifice in the signal passage is used to reduce fluctuations in pressure in the bias spring end of the main spool.

When the machine is shut off, the magnetic field in the solenoid discontinues and a spring forces the pin to seat. At this time, the fluid in the signal channel becomes trapped, but is allowed to bleed to the reservoir through a second orifice in the signal channel.

Manual Bypass

Should an electrical failure occur during operation, the pintle could be seated, closing off the port to the reel drive motor. If this occurs, the operator could not continue to harvest. To overcome this possibility, a manual bypass is incorporated in the valve. To use the manual bypass, proceed as follows:

1. Loosen locknut (Figure 24.48B).
2. With the engine shut off, screw the bypass out of the valve body one turn, then start the machine and check the reel speed. If the reel needs to be adjusted, shut the engine off before making another adjustment. Once the desired reel speed has been set, lock the locknut. The operator will no longer have variable reel speed capability, but at least will be able to harvest until the failure can be corrected. *Caution*: If the manual bypass is opened, the reel will turn at all times that the engine is running, regardless of the separator or feeder switch positions. *Note*: The resistance of the solenoid of the reel drive valve should be 9.6 plus or minus 1.0 ohms.

25

Thermoplastic Injection Molding Machine

A thermoplastic injection molding machine is an excellent example of the troubleshooter's need for a thorough understanding of the hydraulic circuit, its components, and the functional interrelationships of the various elements of a complete molding cycle. (Courtesy of Van Dorn Demag Corporation.)

The molding cycle consists of: 1) closing the mold, 2) injecting a prepared charge of plasticized material into the mold, 3) providing a brief time for the molded part to cure, 4) preparing the next charge of plasticized material, and 5) opening the mold and ejecting the finished part. The combining of all of the control elements leading to a quality finished part requires consistent and precise use of force and motion.

When a system component either gradually or suddenly malfunctions, the troubleshooter should be able to make a quick and efficient evaluation as to which component or components are at fault. Machine downtime and attendant lost production are costly. The ability to quickly and accurately find the fault does not result from guesswork, but from logical thought processes.

Certain system components affect many elements of the molding cycle. One example is the pumps whose output flows are directed to every segment of the complete system. Other components only affect a specific operation in the complete cycle. Another example is the ejection manifold whose only requirement is to control the ejection cycle. The problem then becomes one of how to eliminate those components that are the least likely to be malfunctioning, and how to zero in on those that are the most likely to have failed.

Problems in hydraulic systems generally can be defined as relating to flow, pressure, or direction of flow. They manifest themselves in the form of noise, vibration, heat, or leaks. These manifestations can frequently be detected by the human senses of sight, touch, and hearing. These senses are controlled by our brain, which is capable of analyzing and arriving at judgement decisions. Thinking is the starting point for troubleshooting. Acting follows immediately thereafter when the thought process has reached the mature judgement stage.

Attempting to visualize the hydraulic system and how it functions by looking at the machine first can be self-defeating. The hydraulic components are not visible from a single vantage point. Time can be lost becoming familiar with the physical components of the hydraulic system. This time is better spent reviewing the circuit diagram while relating the trouble symptoms to those components which could be at fault.

The troubleshooter is under pressure to arrive at a solution. Self-confidence gained by keeping a clear mind and approaching a solution in a logical fashion contributes to an early and successful conclusion.

It is possible that the malfunction can be with one of the components that serves the entire system. However, it is also likely to be a localized problem. Having a clear picture of each one of the main subsystems will shorten the time to reach an accurate solution. Identifying those components in the five principle functions of a mold cycle by highlighting the components can assist the troubleshooter in focusing on the problem.

The molding machine builder prepares a chart of the control steps of a complete cycle. It is in the form of a solenoid energization chart. The chart meshes together the required control signals for both the electrical and hydraulic devices to assure that each step of the molding cycle is accounted for. Many different molds will be used in the same machine, each having its own operational characteristics.

For the hydraulic system troubleshooter, this chart is a major source of information to use in zeroing in on the problem. The information provided by the circuit diagram and the solenoid energization chart form the basic tools that are the logical starting point for troubleshooting.

Figures 25.1–25.12 provide a detailed analysis of the steps in a typical molding cycle. Each contains that portion of the circuit which is operative when certain control valve solenoids are energized.

The troubleshooter mentally ties together what is happening within the molding machine during each cycle step, which provides the *first thought* development as to whether or not the malfunction can be logically attributed to those components being activated during that particular step.

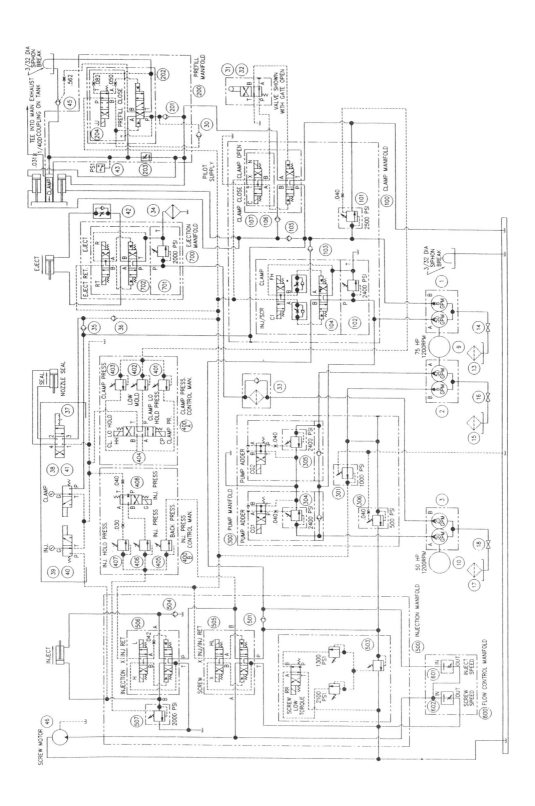

Figure 25.1 Basic hydraulic circuit diagram.

THE CIRCUIT DIAGRAM

The following information is devoted to explaining the operational steps of the molding cycle. The format selected is to accompany each highlighted circuit with a description of what occurs when various solenoids are energized. This format will aid circuit diagram interpretation.

1. Each of 11 circuit drawings (Figures 25.2–25.12 have certain flow path lines highlighted by means of a higher density line. This format quickly identifies those fluid lines that direct specific pump flows to the actuator(s) in use at the time. System pilot lines are identified in the same way. The pilot line design follows the specification data as shown in Chapter 1. Those pilot lines that are used within a pressure relief valve subsystem use a dash–dot format (–•–•–). All three lines "active system flow," "system pilot flow," and "pilot pressure selection flow" appear on each of the 11 circuit drawings and are listed under the title "Legend".

2. The directional control valve solenoids that are energized during each step are highlighted by the addition of an asterisk positioned beside the solenoid.

3. When a directional valve solenoid is energized to provide the required control function, the symbol is modified to depict the flow path through the valve that is taking place at that time.

4. Figure 25.1 provides the complete circuit diagram used in this example. Figures 25.2–25.12 have been modified to simplify the explanation of each of the 15 steps of a molding cycle. The 15 steps of this example were selected only to provide material for the troubleshooting exercises included in this chapter. The modifications made resulted in the removal of many of the circuit symbols of those hydraulic components not required for the molding cycle step being discussed.

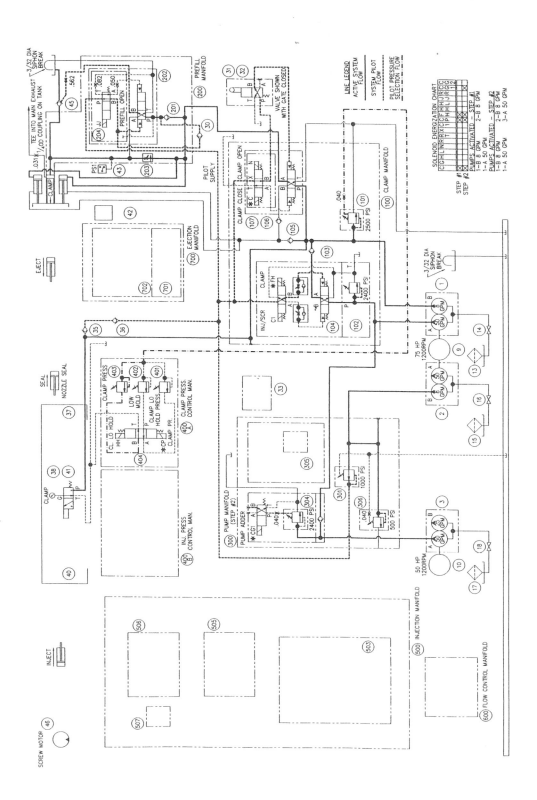

Figure 25.2 Steps 1 and 2: Clamp closes fast.

Clamp Closing

Figure 25.2 should be used to follow steps 1 and 2.

Step 1. Energizing solenoids C, CP, and FH directs flow from pumps #1A and #1B into the main ram and "fast forward" cylinders. Since solenoid JJ (valve 202) is de-energized, it directs pilot flow to the prefill valves (45) pilot line; this opens the valve and allows flow from the overhead reservoir to fill the main ram as it moves towards mold close position.

During this phase, re-generative flow from the main ram and the rod ends of the "fast forward" cylinders joins the flow directed to the cap end of these cylinders.

System clamp pressure is established by energizing solenoid CP. This allows the remote relief control valve (403) to set the main clamp pressure relief valve (101) at the desired clamp pressure. This pressure setting will be lower than the value at which the main relief valve (101) has been set.

At this point, the main ram is moving towards mold close position at a rate provided for by the outputs of pumps 1A and 1B plus the re-generated flow.

Step 2. The same solenoids are energized as in Step 1. Energizing solenoid CG1 closes relief valve 304 diverting flow from pump 3A into the mold close cylinders. The result is a higher closing speed, which helps maximize the production capability of the molding machine by reducing overall cycle time.

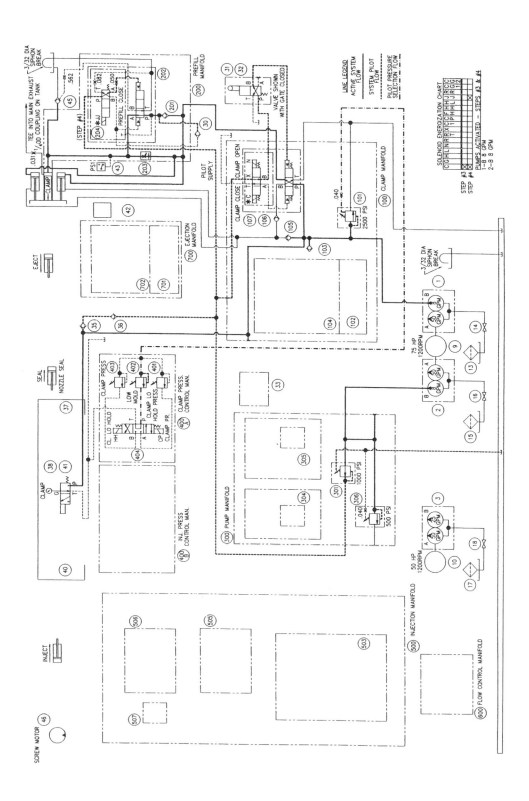

Figure 25.3 Steps 3 and 4: Clamps slow down at low pressure. Prefill valve closes and "mold safe" mode is provided.

Clamp Shut

Figure 25.3 should be used to follow steps 3 and 4: clamp slows down, mold is in safe mode, and prefill valve closes.

Step 3.   Solenoid C remains energized, directing flow from pump 1B resulting in "mold close" travel speed slowing down as pump 1A and 3A are diverted back to tank. Solenoid CP is now deenergized, which allows remote control relief valve 402 to set the main relief valve at low mold pressure. As the mold nears the closed position, the condition called "mold safe" is now in effect.

Should any foreign material be trapped between the mold halves, the low closing pressure will protect the mold from damage. Once this is ascertained, Step 4 is activated. Regeneration is still in effect.

Step 4.   Solenoid JJ is energized, shifting the second stage of valve 202 and allowing prefill valve to close as its pilot pressure is diverted to the tank. This stops the regenerative cycle as the mold is now closed.

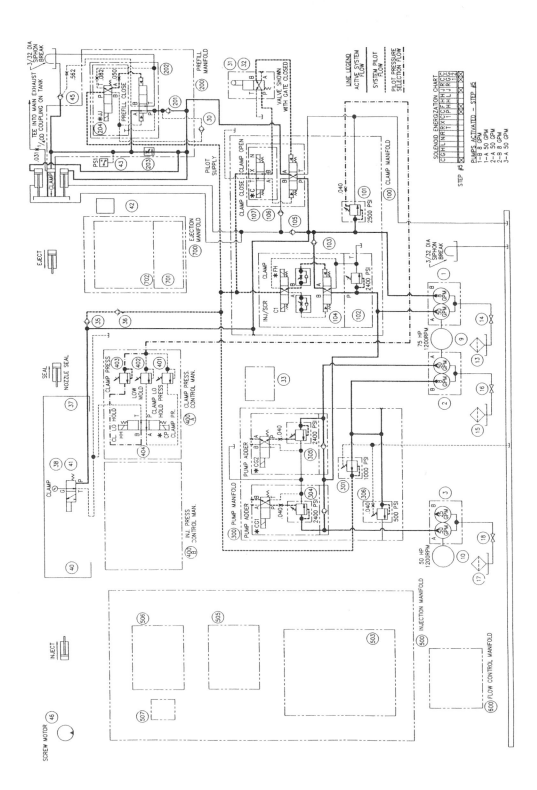

Figure 25.4 Step 5: Clamp pressure build-up.

Clamp Pressure Build-Up

Figure 25.4 should be used to follow step 5.

Step 5. Solenoid JJ remains energized, thus closing the prefill valve. Solenoid CP is energized, activating the remote relief control valve (403) and setting main clamp relief valve (101) at "clamp pressure." Solenoid C remains energized, directing pump 1B flow to the clamp circuit. Solenoids CG1 and CG2 are energized to close relief valves 304 and 305; this diverts flow from pumps 2A and 3A into the clamp circuit via valve 104 as solenoid FH is energized.

The purpose of the "clamp pressure build-up" is to generate the desired rate of flow to build pressure very rapidly in the clamp circuit. This is the same objective as in "clamp closes fast" cycle. Both increase machine productivity by reducing the amount of time to perform the step.

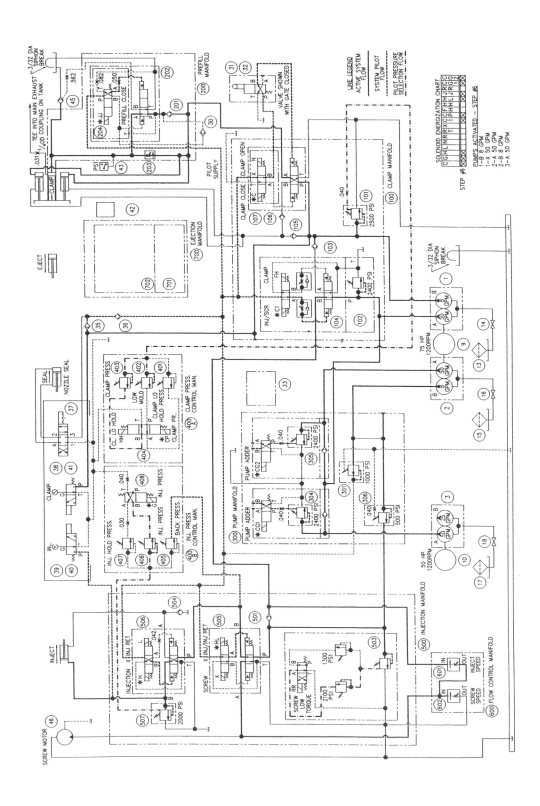

Figure 25.5 Step 6: Injection boost.

Injection Boost

Figure 25.5 outlines step 6.

Step 6. Solenoid C remains energized, directing flow from pump 1B into the clamp circuit to maintain clamp pressure. Solenoid CP remains energized, resulting in "clamp pressure" being maintained in clamp circuit. Solenoid JJ remains energized, keeping the prefill valve closed. Solenoids CG1 and CG2 remain energized, providing flow from pumps 2A and 3A into the "injection boost" cylinder. Solenoids C1, H, and HL are energized to direct flow from pumps 1A, 2A, and 3A into the cap end of the injection cylinder. Solenoid G is energized to activate the remote relief valve (406) to set the main injection relief valve (507) at the desired injection pressure.

The purpose of the "injection boost" circuit is to rapidly move the plasticized material in storage in the mold puddle ahead of the mold into the mold. Flow control valve 601 provides a means of adjusting the speed of the injection boost cylinder. This adjustment provides the flexibility needed to compensate for different mold cavity sizes. The flow control operates as a bleed-off type, returning flow not required back to the tank via valve 505.

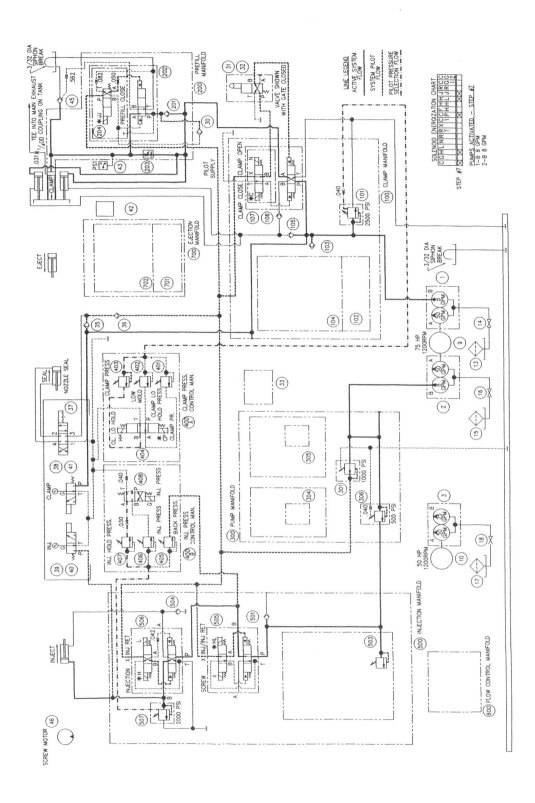

Figure 25.6 Step 7: Injection hold.

Injection Hold

Figure 25.6 outlines step 7.

Step 7. Solenoid C remains energized to maintain flow from pump 1B into the clamp circuit. Solenoid CP remains energized to maintain "clamp pressure". Solenoid JJ remains energized, keeping the prefill valve closed. Solenoids H and HL remain energized to direct flow from pump 2B into the injection cylinder.

The purpose for "injection hold" is to provide a period of transition from the injection phase to the screw rotate phase. This is the first time that pump 2B is providing flow into the system for a reason other than system pilot pressure.

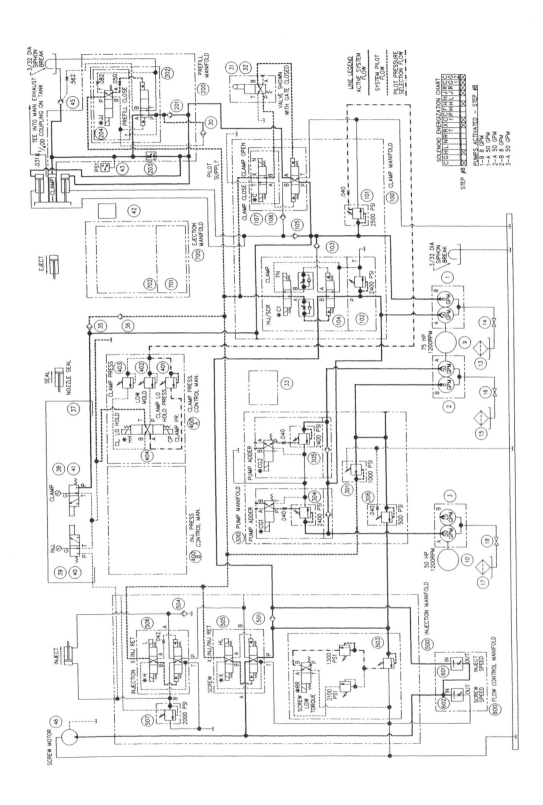

Figure 25.7   Step 8: Screw rotate.

Screw Rotate

Figure 25.7 outlines step 8.

Step 8.   Solenoid C remains energized to maintain flow from pump 1B into the clamp circuit. Solenoid HH is energized, activating the remote relief valve (401) and setting the main clamp relief valve (101) at "clamp low hold pressure." Solenoid JJ remains energized, keeping the prefill valve closed. Solenoid C1 is energized, directing flow from pumps 1A, 2A, and 3A into the screw rotating motor. Solenoids CG1 and CG2 are energized to divert flows from pumps 2A and 3A into the screw rotate circuit. Solenoid X is energized to direct pump flows into screw motor. Solenoid H is energized to allow oil in cap end of inject cylinder to return to tank. Solenoid RR is energized to set relief valve 503 at "screw low torque" value of 1300 psi.

The purpose of the "screw rotate" is to replenish the supply of plasticized plastic in the mold puddle cavity to prepare for the next injection boost operation to make another molded part. Note that the cap end of the injection cylinder is open to the tank so that the slide mechanism carrying the plasticizer screw can move backward as the mold puddle cavity becomes filled with plasticized plastic. The screw rotating RPM can be varied by adjustment of the flow control valve (602). This is the bleed-off application of a flow control valve. This is the second time pump 2B is providing flow into the system other than for pilot pressure use.

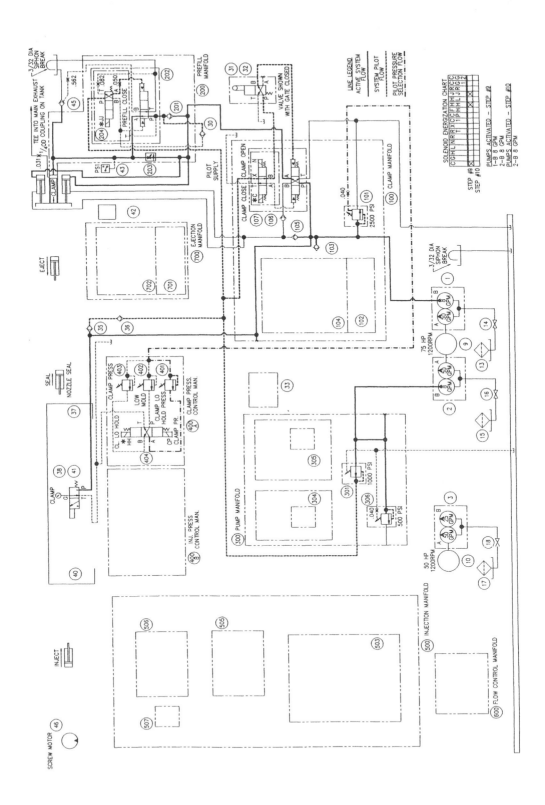

Figure 25.8 Steps 9 and 10: Cure time and clamp decompresses.

Cure Time/Clamp Decompresses

Figure 25.8 outlines steps 9 and 10.

Step 9. Solenoid C remains energized to maintain flow from pump 1B into clamp circuit. Solenoid HH remains energized activating the remote relief valve (401) at "clamp low hold pressure."

Solenoid JJ remains energized keeping prefill valve closed. The purpose of the "cure time" is to allow the plasticized plastic now in the mold time to cure and become strong enough to be ejected from the mold after it is opened.

Step 10. All solenoids are de-energized. This is the decompression phase when oil passing through a 1/32 inch orifice just outside of the main ram provides the means for clamp pressure to start to decay.

The purpose of step 10 is to eliminate system shock by bleeding down system pressure prior to unclamp phase. Opening the prefill valve further decompresses the system.

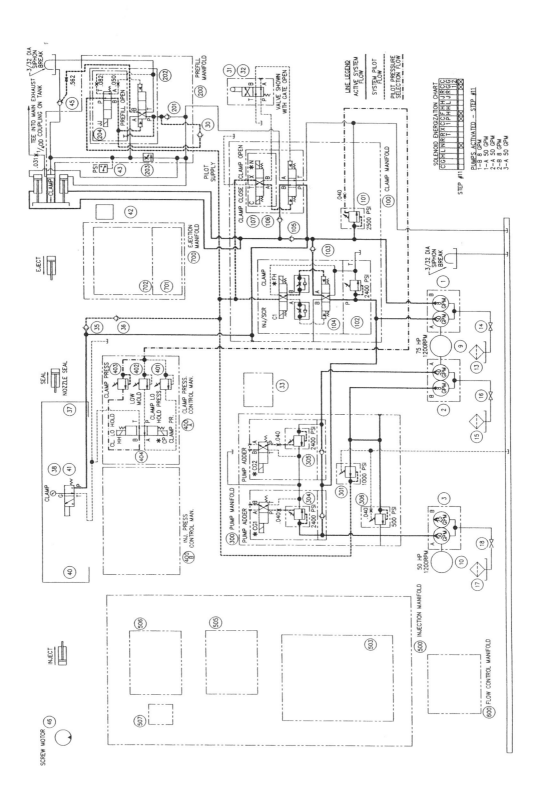

Figure 25.9 Step 11: Clamp opens fast.

Clamp Open

Figure 25.9 outlines clamp "fast open" and then the stopping of the clamp action as the core pull is made, followed by continuing of the "fast open" movement.

Step 11. Solenoids N and FH are energized to direct flow from pumps 1B, 1A, 2A, and 3A into the rod end of ram retract cylinders. Solenoid CP is energized activating the remote relief valve (403) and setting the main clamp relief valve (101) at "clamp pressure". Solenoids CG1 and CG2 are energized to divert flow from pumps 2A and 3A into the main ram retract cylinders.

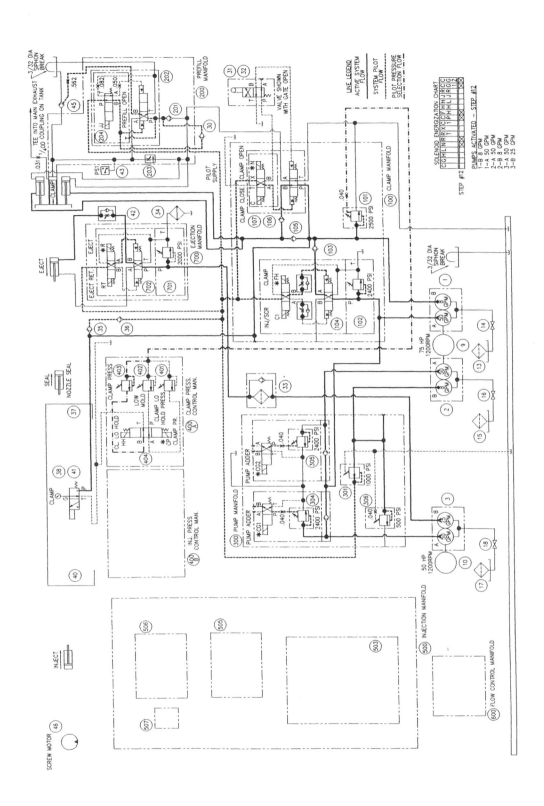

Figure 25.10 Step 12: Eject forward.

Mold Eject

Figure 25.10 outlines the continuing of the clamp opening as "fast open," and the activation of the "eject forward" cylinder.

Step 12. Solenoid R is energized, directing flow from pump 3B to the cap end of the eject cylinder to eject the part from the mold. At the same time, solenoids N, FH, CG1, and CG2 remain energized, continuing to direct flow from pumps 1B, 1A, 2A, and 3A to the main ram retract cylinders. Solenoid CP remains energized to provide "clamp pressure" setting of relief valve 101.

Pilot pressure continues to hold prefill valve open for ram discharge flow to return to tank. The purpose of step 12 is to maintain clamp opening at "fast open" speed and eject the part from the mold.

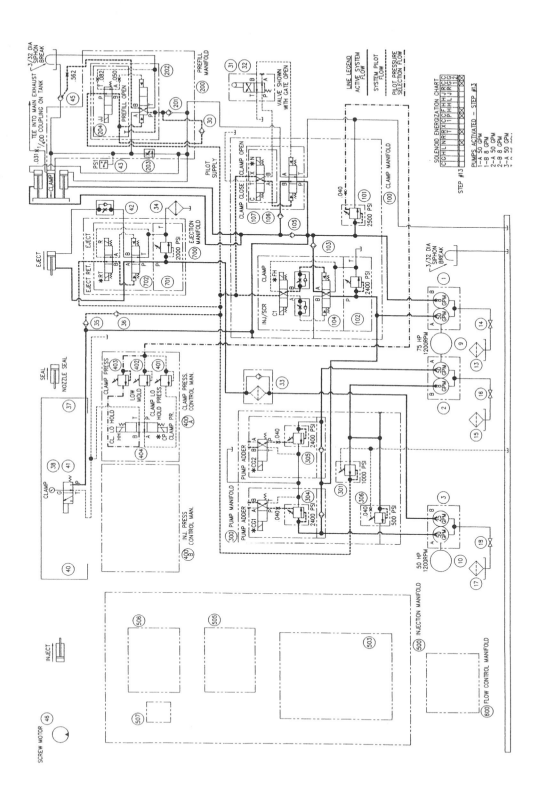

Figure 25.11 Step 13: Ejector retracts.

Retraction of the Ejector Cylinder

Figure 25.11 outlines the continuing of the clamp opening at "fast open" and the retraction of the ejector cylinder.

Step 13. Solenoid RT is energized directing flow from pump 3B to the rod end of the ejector cylinder. At the same time, solenoids N, FH, CG1, and CG2 are continuing to direct flow to the main ram retract cylinders. Solenoid CP remains energized to provide "clamp pressure" setting of relief valve 101.

The purpose of step 13 is to maintain clamp opening at "fast open" speed and to retract the eject cylinder.

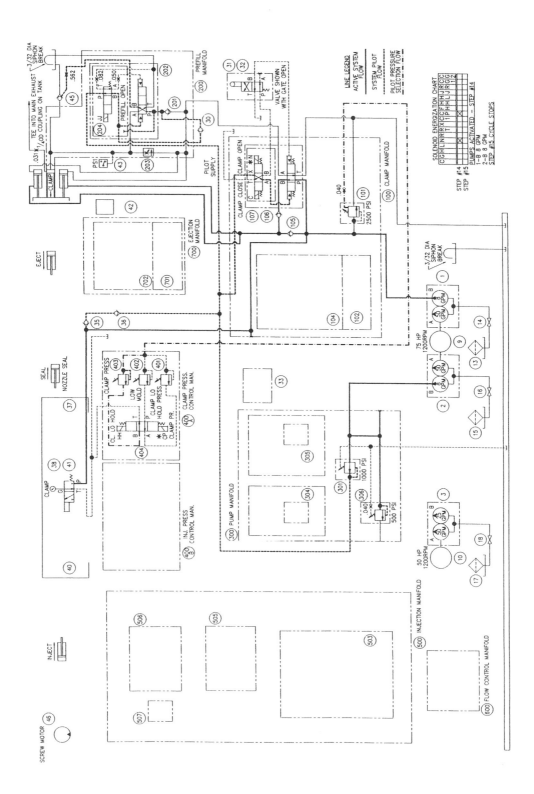

Figure 25.12 Steps 14 and 15: Clamp slow down and stop.

Clamp Slow Down and Stop

Figure 25.12 outlines "clamp slow down" and "stop" and the end of the mold cycle.

Step 14.   Solenoid N remains energized continuing to direct flow from pump 1B resulting in a slow down of the clamp open cycle. Solenoid CP remains energized to provide "clamp pressure" setting of relief valve #101. Pilot pressure holds prefill valve open for ram discharge flow to return to tank.

Step 15.   All solenoids are de-energized, signalling the end of the mold cycle. All pump flows are directed to the tank. The completed part is removed and the molding machine is ready for the next cycle.

INVESTIGATIVE PLANS

The remainder of this chapter focuses on the thought process of establishing an investigative plan to quickly and efficiently find the problem in the hydraulic system. The objective is to motivate the troubleshooter to apply logical mental analysis prior to taking physical action. Three exercises follow—each is based on a conceivable problem with a hydraulic component.

Exercise 1

The molding machine operator reported that following removal of the molded part the machine would not respond to the command signal to start the next molding cycle. The operator said that nothing to his knowledge had happened. Prior to this new malfunction the machine had performed properly. No slow down or other outward manifestation as to what was wrong had taken place. The clamp cycle simply would not start.

Having learned this from talking with the machine operator, the troubleshooter should first review the circuit diagram and solenoid energization chart. Two conditions could be at fault:

Lack of flow to the "clamp closes" cylinders
Lack of pressure to do the work

The logical step is to consider which component (or components) is most likely to be at fault. This includes preparation of a list of those components that are being employed during that step in the cycle. The list should contain:

1. Pumps (flow)
2. Relief valves (pressure)
3. Clamp cylinders (flow and pressure)
4. Directional valves (flow)
5. Pilot pressure pump (flow and pressure)
6. "Open" safety gate valve (flow)

The next step is to prioritize the problem possibilities so that an action plan can be developed.

The troubleshooter should turn to the circuit and the solenoid chart for ease of visualizing the components in the clamp circuit. To appreciate the value of these tools, refer to Figure 25.2 and step 1. Table 25.1 lists initial thoughts about the likelihood of various components being the cause of the fault, and the priority for troubleshooting each. Logic indicates to check the high priority items first.

Both items 4 and 6 carry a number 1 priority. These items could provide a significant return towards finding the fault. The reason is that in order for flow from pumps 1A and/or 1B to reach the clamp cylinders, valves 107 and 31 must be actuated. A problem with either one would stop this from happening.

Valve 31 is mechanically operated by the safety gate. A visual inspection of the valve spool can provide a preliminary indication as to whether or not this is the problem.

This example is based on solenoid C on valve 107 having burnt out.

NOTE: Had only solenoid FH on valve 104 burnt out, the clamp would have moved forward at a slow speed equal to the output flow of pump 1B. This would not have met the problem criteria that "no clamp forward motion had occurred."

Table 25.1 Possible Problem Components in Exercise 1

Item	Component	Thought process	Priority rating
1	Pumps (1A, 1B)	Not likely a pump. The previous cycle completed satisfactorily. No noise from pumps	3
2	Relief valve	A possibility due to contamination and subsequent piston/body bore scoring, which would freeze the piston in open position.	2
3	Clamp cylinders	Not likely. Cylinder failure due to piston packing wear would be gradual, but this did not occur.	3
4	Directional valve (107)	A possibility. Solenoid burn out would fill the condition associated with this fault.	1
5	Pilot pressure (pump 2B)	Not likely. The pump did not display pressure failure symptoms	3
6	"Open" safety gate valve (31)	A possibility. If the valve was not closed to direct pilot pressure flow to second stage of valve 107, it in turn would not direct flow to the clamp circuit.	1

If the number 1 priority item had not been at fault then the troubleshooter would go on to number 2 and then number 3 items. If all three priorities were exhausted without finding a solution, then a new priority list should be made and acted on.

The exercise just concluded incorporated a simple fault of solenoid burnout. This exercise and the ones that follow are designed to encourage the troubleshooter to think first, then troubleshoot.

Exercise 2

The machine operator noted that the finished parts were varying in their degree of completeness from one molding cycle to the next. Obviously a machine malfunction was developing. It was not long before the operator had to shut down the machine due to unsatisfactory parts. The operator sensed that the injection cylinder was varying in its "inject forward" speed.

Following a discussion with the operator, the troubleshooter concluded that the problem was one of inconsistent movement of the "inject cylinder." The problem at hand was to determine if the fault related to inconsistent flow or pressure or both.

The troubleshooter elected to activate the inject system pressure gage. He noted that when an incomplete part was made, there was some drop off in system pressure, but it was not substantial.

Again, the logical step here is to identify what components relating to the "injection boost" cycle are involved. The list should include:

1. Pumps (flow)
2. Relief valves (pressure)
3. Inject cylinder (flow/pressure)
4. Inject flow control valve (flow)

Table 25.2   Possible Problem Components in Exercise 2

Item	Component	Thought process	Priority rating
1	Pumps (1A, 2A, 2B, 3A, 3B)	Not likely a pump. The previous cycles using the same pumps did not experience similar problems.	3
2	Relief valves (304, 305, 507, 407)	A possibility due to contamination and/or scoring of piston/body bore, causing erratic relief action.	2
3	Injection cylinder	A possibility due to piston packing wear.	1
4	Inject flow control valve 601	A possibility due to contamination scoring the hydrostat causing erratic action	1

To prepare a list of potentially faulty components (Table 25.2) and to prioritize them before taking investigative action, refer to figure 25.5 and step 6. Note that Solenoids C, G, H, C1, CP, HL, JJ, CG1 and CG2 are energized.

Because there was no evidence of external system fluid leakage, the fault must be internal leakage. When this occurs, the result is generation of heat as high pressure oil is forced through a small orifice. Frequently the heat is discernable by touch; you can use your hand to detect the difference in temperature entering and leaving the suspected leak point.

The decision was made to check the cylinder first, because the fluid exiting the rod end seemed to be at a higher temperature than that entering the cap end. A check of the cylinder was made by bottoming the piston against the rod end, disconnecting the piping at the rod end, and starting the pump. Some bypass leakage was found, but it did not seem severe enough to have caused the problem itself.

The next candidate on the priority list was the inject flow control valve. The valve was removed and inspected. Scoring of the hydrostat and its body bore was found to be sufficient to cause erratic flow control action resulting in speed variation by the injection cylinder. It is the function of the flow control valve hydrostat (compensator) to present a constant pressure drop to the variable orifice no matter what flow setting is established. If this constant pressure drop is not maintained, the flow control valve becomes a simple orifice allowing flow to vary with varying pressure.

In practice, there are fine shades of analysis leading to prioritizing of fault possibilities. Experience soon allows one to develop a better sense of which component to select for early inspection when several candidates carry equal priority ratings.

Exercise 3

The molding machine was at that part of the molding cycle where the clamp fast open motion was scheduled to start. The operator noticed that it was sluggish and not moving at the normal fast open rate.

The troubleshooter's first impression was that the problem was flow. He also recognized that lack of pressure to accelerate the mass of the mold slide and associated machine parts could be a factor.

The list of hydraulic system components that are candidates for review should include:

1. Pumps (flow)
2. Relief valves (pressure)
3. Fast open cylinders (flow)

To prepare a list of potentially faulty components (Table 25.3) and to prioritize them, refer to Figure 25.9 and step 11. Note that solenoids N, CP, FH, CG1, and CG2 are energized.

Since the pumps were operating properly on the previous cycle and did not show evidence of any sudden pumping problems, and since it was unlikely that the fast open cylinder(s) would develop a significant enough piston seal bypass leak in a matter of seconds between cycles, the slow down of fast open clamp rate must be due to internal leakage.

With an open relief valve causing partial flow to be returned to tank being a good candidate for initial investigation, the question is which relief valve should be checked first. A review of the circuit shows that if valves 101 and 102 were open or partially open, the problem would occur. Of these two, which should be selected? The odds are in favor of starting with valve 101, because it is subject to the same failure modes (scoring of the main spool or seat, contamination in the balance hole of the main spool, and contamination in the pilot stage) as valve 102, but also to the conditions of its remote control relief valve (403), which could have similar problems.

Having selected valve 101 for investigation, the logical next step is to isolate valves 101 and 403 from each other and test each individually. This can be accomplished by disconnecting the pilot line from valve 403 and inserting a plug fitting in the remote control port of valve 101. Then back off the pressure adjustment screw of valve 101 to attain a low system pressure when the pump is re-activated.

The purpose of this test is to witness the action of the valve. Start up the hydraulic system and position the actuator so that system pressure can be achieved. Next, build system pressure by adjusting the valve. If system pressure can be achieved, then move on to check the remote control 403.

In this example, system pressure could not be achieved. The next action is to disassemble valve 101 to find the malfunction.

As it turned out the balance hole had finally reached the point where insufficient flow through the orifice brought on the malfunction. The resulting low system pressure did not

Table 25.3   Possible Problem Components in Exercise 3

Item	Component	Thought process	Priority rating
1	Pumps (1A, 1B, 2A, 3A)	Not likely a pump. The previous cycle of the fast open used the same pumps but did not experience a problem; there is no evidence of a sudden failure.	3
2	Relief valves (101, 102, 403)	A possibility. Although seemingly unlikely, these valves are likely because no problem occurred previously but now there is a problem in step 11.	1
3	Fast open cylinders	It is not likely that the cylinder(s) would develop an internal leak so suddenly.	2

provide enough force from the fast open cylinders to accelerate and operate the cylinders at a normal fast open rate.

SUMMARY

The exercises above are simplistic by design. The objective of this chapter is to highlight the necessary thought processes and encourage troubleshooters to make use of their most valuable tools: the circuit diagram and the solenoid energization chart.

It is the writers' view that the technique of hydraulic system troubleshooting needs to be substantially elevated from the "cut and try" method of arbitrary selection of a component so enshrined by the "remove and replace" syndrome to a "think first" attitude. With this change of attitude, the next step is to become comfortable with the hydraulic circuit diagram and the graphic symbols that describe the function of each component.

To be able to find the source of the malfunction quickly, troubleshooters must understand what the machine is supposed to do. Knowledge of what each step in the complete cycle entails is the key to fast and efficient results.

The circuit diagram is a concise road map that graphically demonstrates the path by which system fluid flows. It shows the control components that decide the flow characteristics, the direction of flow, and the pressure levels. The final destination is the actuators that do the work.

It is much simpler to study and review these mileposts by using the circuit diagram; rather than try to decipher what is happening by looking at the machine as a whole.

Using the solenoid energization chart provides the tools for breaking down the complete system into segmented parts. This allows the troubleshooter to focus on a specific set of circumstances at a specific period of time.

There are particular benefits in freeing the mind to work on portions of a problem as compared with trying to take on the complexity of the whole.

Every person has an identity of his or her own. We all work in different ways. However, when we are ultimately faced with a complex problem, thinking first and *then* acting will usually pay the greatest dividend.

Index